Understanding
Digital Logic Circuits

Bob Middleton has been a professional free-lance technical writer in the electronics field for many years. His numerous books and magazine articles have proven invaluable to students, technicians, and engineers, because they have all been based on his own practical experience. His home workshop is filled with the latest test equipment, receivers, and other equipment that he uses to check out every detail in preparing his books.

Other Sams books by Mr. Middleton include *101 Ways to Use Your VOM, TVM, & DVM, Radio Training Manual, Know Your Oscilloscope, Troubleshooting With the Oscilloscope, Effectively Using the Oscilloscope,* and many others.

Understanding
Digital Logic Circuits

by

Robert G. Middleton

Howard W. Sams & Co., Inc.
4300 WEST 62ND ST. INDIANAPOLIS, INDIANA 46268 USA

Preface

This is a working handbook for the service technician who is now engaged in radio, television, audio, or related areas of electronic troubleshooting and repair and who would like to expand his or her expertise in the digital electronics field "painlessly." This book is designed to prepare the reader to cope successfully with digital-logic circuitry in scanner-monitor radios, two-way radios, home appliances, television digital tuners, logic-controlled tape decks, video cassette recorders, digital clocks, video games, digital hobby equipment, transmitter control equipment, and digital test instruments. The digital revolution confronts all of us, and we must understand this new technology if we are to survive in the electronics field.

This practical handbook starts with an overview of the "anatomy" of digital logic diagrams. Commercial IC packages of AND, OR, XOR, NAND, NOR, XNOR, and AND-OR-invert gates, buffers, and inverters are described and illustrated, with examples of applications. Basic combinatorial-logic configurations are explained. Fundamental relations of digital-logic diagrams to preliminary troubleshooting processes are noted. Simple digital-logic test equipment is described.

In the second chapter, additional logic gates and circuit operation are described and illustrated. Basic adders and subtracters are explained in the third chapter. It is shown how an adder is operated to subtract binary numbers. The new majority gates are introduced in Chapter 4, with an introductory coverage of CMOS logic. The essentials of flip-flops and clock circuitry are described and illustrated in Chapter 5. The sixth chapter covers additional flip-flops and monostable circuitry.

Fundamentals of latch-type, fall-through, and shift registers are discussed in Chapter 7. The related area of digital counters is covered in the eighth chapter and is continued in Chapter 9 with discussion and examples of frequency dividers. Practical applications and troubleshooting notes are included. The tenth chapter introduces the reader to various types of encoders and decoders, with an explanation of code conversion. Parity generator/checkers and interfaces are described and illustrated in Chapter 11, with notes on interfacing TTL circuitry to CMOS, and CMOS circuitry to TTL.

Other forms of code converters, multiplexers, demultiplexers, and comparators are discussed in Chapter 12, with examples of practical applications. The coverage proceeds to programmable counters and special shift registers in Chapter 13. Look-ahead–carry adders and binary-coded–decimal adders are described and illustrated in Chapter 14; this chapter includes explanations of binary multiplication and division. The important topic of shift-register memories, random-access memories, and read-only memories is covered in Chapter 15, with attention to both bipolar and CMOS circuitry.

Additional memories, such as PROMs, EPROMs, and related devices, are discussed in Chapter 16. Additional bipolar memories and mass dynamic-MOS memories are explained in Chapter 17. The practical topic of digital voltmeter circuitry is covered in Chapter 18. Although often neglected, the important topic of transmission lines in digital systems deserves close attention; it is explained in Chapter 19.

Electronics technology is becoming increasingly sophisticated. We technicians must take this progress in stride if we are to remain competitive. Unless we can read a digital logic diagram as well as we can read a television circuit diagram, it will become increasingly difficult in the future for us to service consumer electronic products properly and profitably. In preparing this handbook, I have recognized this growing need and have made a dedicated effort to meet it. I may add that "reinforced learning" gained by working with digital equipment at the bench will be much more valuable to you than the knowledge you can gain from merely reading this or any book.

ROBERT G. MIDDLETON

Contents

CHAPTER 1

CHAPTER 2

CHAPTER 3

CHAPTER 4

Configurations — I²L Gates — Totem-Pole Troubleshooting — MOSFET Technology

CHAPTER 5

CHAPTER 6

CHAPTER 7

CHAPTER 8

CHAPTER 9

shooting Notes—Digital Clock—Counters in Home Appliances—Frequency Division in a Digital Color-Bar Generator

CHAPTER 10

CHAPTER 11

CHAPTER 12

CHAPTER 13

CHAPTER 14

CHAPTER 15

CHAPTER 16

CHAPTER 17

CHAPTER 18

CHAPTER 19

Chapter
1

The Anatomy of Digital Logic Diagrams

Technicians who have had experience only with analog circuitry may suppose that digital logic diagrams are very difficult to understand. On the contrary, digital logic diagrams can be mastered easily by the technician who starts at the beginning and proceeds step-by-step.

GATES

Most digital logic circuits are arrangements of gates in various "disguises." A *gate* is an electronic switch that contains one or more transistors (in most cases). The gate either is "on" or it is "off." In other words, a gate transistor is either saturated or it is cut off. One of the basic gates is the two-input AND gate (Fig. 1-1). This AND gate may be considered functionally as two spst switches (electronic switches) connected in series.

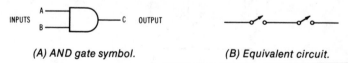

INPUTS A B — C OUTPUT

(A) AND gate symbol.   *(B) Equivalent circuit.*

Fig. 1-1.   Basic two-input AND gate.

In Fig. 1-2, it can be seen that an output from the AND gate will be obtained if both switch A and switch B are closed simultaneously. If either (or both) of the switches is open, no output will be obtained from the AND gate. If input A is at ground potential, its corresponding switch is open (transistor cut off). If input A is at a +2-volt potential, its corresponding switch is

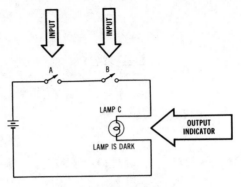

(A) Both switches open.

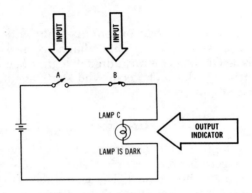

(B) One switch closed.

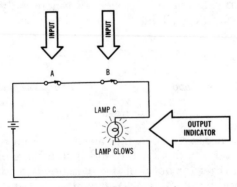

(C) Both switches closed.

Fig. 1-2. Operation of basic two-input AND gate.

closed (transistor turned on and saturated). Ground potential is called the logic-low state; this state may be referred to as "false," or it may be symbolized by L or 0. A +2-volt (or perhaps higher) potential is called the logic-high state; this state may be referred to as "true," or it may be symbolized by H or 1. A truth table lists all possible input states, with the resulting output states (see Fig. 1-3).

INPUTS		OUTPUT
A	B	C
0	0	0
0	1	0
1	0	0
1	1	1

Fig. 1-3. Truth table for two-input AND gate.

BASIC INTEGRATED CIRCUITS

Most of the AND gates that you will encounter are in integrated circuit (IC) form (Fig. 1-4). Each IC package will usually contain more than one AND gate, as shown in Fig. 1-5. The device represented in Fig. 1-5 is a type 9N08 quad two-input AND gate; "quad" denotes that the package contains four individual gates. This type of diagram is often termed a *connection diagram*, or a *logic and connection diagram*. Observe that IC pins

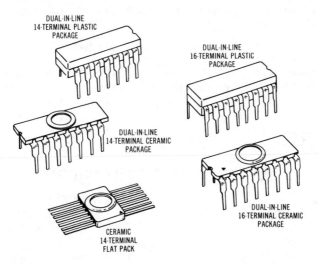

Fig. 1-4. Appearance of digital integrated circuit packages.

13

1 and 2 are the input terminals for an AND gate and that pin 3 is the output terminal for the same gate. Pins 4 and 5 are the input terminals for another AND gate, and pin 6 is the output terminal for that gate.

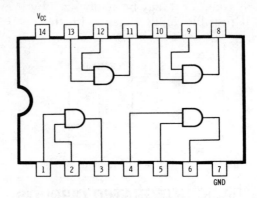

Fig. 1-5. Package pinout diagram for a quad two-input AND gate.

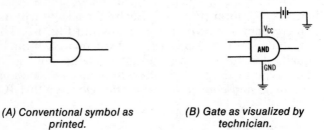

(A) Conventional symbol as printed.   *(B) Gate as visualized by technician.*

Fig. 1-6. Visualization of AND gate symbol.

Technicians refer to IC terminal identifications as the *package pinout*. Do not overlook the V_{CC} terminal (pin 14) and the ground terminal (pin 7). The gates will not operate unless a power source is connected between the V_{CC} and ground pins. In other words, the logic symbol for an AND gate is visualized by the technician as shown in Fig. 1-6B.

DIGITAL-LOGIC TEST PROBE

Chart 1-1 outlines the use of digital-logic diagrams in troubleshooting procedures. Note that checks are made of the input/output relations of suspected devices (such as AND gates), with reference to their truth tables.

One of the most useful units of digital-logic test equipment is a *logic probe*. A very simple type of logic probe is shown in Fig.

Chart 1-1. Use of Digital Logic Diagrams in Troubleshooting Procedures

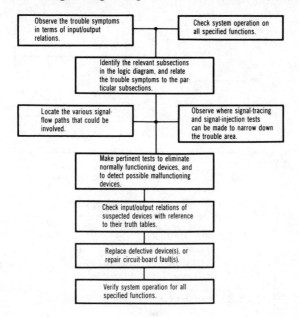

Observe the trouble symptoms in terms of input/output relations.

Check system operation on all specified functions.

Identify the relevant subsections in the logic diagram, and relate the trouble symptoms to the particular subsections.

Locate the various signal-flow paths that could be involved.

Observe where signal-tracing and signal-injection tests can be made to narrow down the trouble area.

Make pertinent tests to eliminate normally functioning devices, and to detect possible malfunctioning devices.

Check input/output relations of suspected devices with reference to their truth tables.

Replace defective device(s), or repair circuit-board fault(s).

Verify system operation for all specified functions.

1-7. It does not require a battery or power supply, because it is connected directly to the equipment being checked (Fig. 1-8). The probe can be built into a plastic tube with the LED mounted at one end and a nail used as a probe at the other end. See the parts list in Table 1-1.

Connect the positive lead from the probe to a positive 5-volt dc terminal of the equipment to be checked and the negative lead to a ground terminal. Then touch the tip of the probe to the various input and output terminals of the device to determine

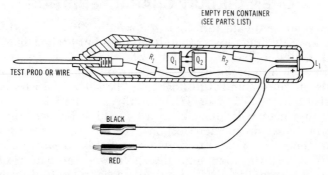

Fig. 1-7. A simple digital-logic test probe. (*Courtesy GC Electronics*)

15

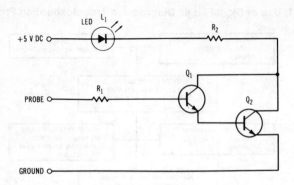

Fig. 1-8. Circuit diagram of logic probe.

Table 1-1. Parts List for Logic Probe

SYMBOL	QTY.	DESCRIPTION	CALECTRO CAT. NO.*
Q1, Q2	2	Npn Silicon Transistor	K4-507
R1	1	27,000-ohm, 1/2-watt resistor	B1-401
R2	1	150-ohm, 1/2-watt resistor	B1-374
L1	1	LED	K4-559
	1	Plastic tube from Pen Oiler, GC Cat. No. 984*	
	2	Jumper Wires	J4-650

*Or equivalent

whether they are on or off (whether the given terminal is logic-high or logic-low). If the indicator LED in the probe glows, it indicates that the terminal is in an on condition; if the indicator does not glow, it indicates that the terminal is in an off condition.

OPEN-CIRCUIT CHARACTERISTICS

Since an AND-gate terminal is normally either logic-high or logic-low, a voltage lower than logic-high but higher than logic-low is called a "bad level." With reference to Fig. 1-9, an open circuit such as a break in a printed-circuit conductor causes the inputs of gates 2 and 3 to float to a "bad level." In transistor-transistor logic (TTL, or T²L) circuitry, a "bad level" at the input to a gate will "look like" a logic-high level to the gate. This perhaps unexpected response results because a floating input generally picks up stray noise fields; occasional noise spikes exceed the logic-high threshold, and the gate responds just as if a logic-high digital pulse had been applied to the input.

Again with reference to Fig. 1-9, a simple logic probe applied at point B will respond to the "bad level" as if it were a logic-low level. (A high-amplitude noise spike is very narrow and does not produce a visible LED indication.) However, when the logic probe is moved to point A, the LED glows and indicates that this is a logic-high level. The inputs of the first gate must be held logic-high during this test; for example, both inputs could be tied to some logic-high point in the system.

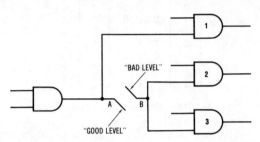

Fig. 1-9.   Effect of an open signal path external to the ICs. (*Courtesy Hewlett-Packard*)

SOME MORE AND GATES

Many of the AND gates encountered by the digital technician have from three to six inputs (Fig. 1-10). Like a two-input AND gate, a multiple-input AND gate has a logic-high output only

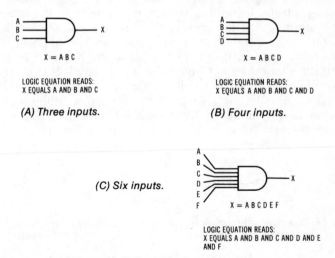

Fig. 1-10. Examples of AND gates with more than two inputs.

A	B	C	D	X
0	0	0	0	0
1	0	0	0	0
0	1	0	0	0
0	0	1	0	0
0	0	0	1	0
1	1	0	0	0
0	1	1	0	0
0	0	1	1	0
1	0	1	0	0
0	1	0	1	0
1	0	0	1	0
1	1	1	0	0
0	1	1	1	0
1	0	1	1	0
1	1	0	1	0
1	1	1	1	1

Fig. 1-11. Truth table for a four-input AND gate.

when all of its inputs are simultaneously logic-high. This is illustrated in Fig. 1-11 for a four-input AND gate.

An AND gate is said to perform the AND *function* when all of its inputs are simultaneously driven logic-high. An AND gate must have at least two inputs, or it cannot perform the AND function. Therefore, it may come as a surprise to learn that in some logic diagrams you may encounter an AND gate that has only one input! To understand how an AND gate operates with a single input, refer to Fig. 1-12. Here, a two-input AND gate is used with both of its inputs tied together. Both inputs will go logic-high if a logic-high voltage is applied; on the other hand, both inputs will go logic-low if ground potential is applied. Therefore, a single-input AND gate does not perform the AND function; instead, it operates as a *buffer*. A buffer is a simple amplifier that does not invert the input signal. It is used chiefly to isolate one digital circuit section from another digital circuit section. A buffer is also used to step up a subnormal digital signal to the standard logic-high level. Logic diagrams may show the symbol as in Fig. 1-12A or as in Fig. 1-12B.

(A) Two-input gate connected for operation with one input.

(B) Equivalent logic symbol used in some diagrams.

Fig. 1-12. An AND gate with only one input.

18

There is a practical reason for the occasional use of AND gates as buffers, instead of using a device that has been designed specifically as a buffer. When a digital circuit is designed, it may happen that there are one or two AND gates "left over" in an IC package. Then, in case the designer has the need for a buffer in the configuration, he will save production costs by connecting one of these left-over AND gates as a buffer (single-input gate).

LOGIC EQUATIONS

Every logic *function* has a corresponding logic *equation*. For example, a two-input AND gate has the logic equation $X=AB$, or $X=A \cdot B$, wherein X denotes the output logic state and A and B denote the input logic states. This logic equation is read, "X equals A AND B." A six-input AND gate has the logic equation $X=ABCDEF$, or $X=A \cdot B \cdot C \cdot D \cdot E \cdot F$; this logic equation is read, "X equals A AND B AND C AND D AND E AND F." It is evident that a logic equation provides a compact summary of the truth table for a gate.

1 and 0 LOGIC STATES

Note carefully that the 1 state does not necessarily imply more power, more energy, or a higher potential than the 0 state; however, we have assumed in the foregoing discussions and examples that the 1 state is the more positive signal level. Although this is often true in practice, it would be an error to assume that it is true for all digital circuits.

With reference to Fig. 1-13, observe that manufacturers of gates specify the AND-gate characteristics in terms of a logic equation; the equations in Fig. 1-13 are also stipulated to be *positive-logic* expressions. Positive logic basically means that a positive supply voltage is connected to the V_{cc} terminal, and that the logic-1 level is more positive than the logic-0 level. This is characteristic of TTL gates, examples of which are shown in Fig. 1-13.

Fig. 1-14 is a schematic diagram of the internal circuit of one four-input AND gate. For most purposes, logic symbols are used instead of schematic diagrams because of the comparative simplicity of a logic diagram. If schematic diagrams were used to represent elaborate digital systems, the network would have a highly confusing appearance, and the diagram would be very difficult to read.

19

DIP (TOP VIEW)

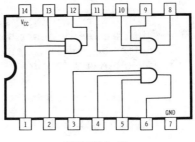

POSITIVE LOGIC: Y = ABC

(A) Type 9H10.

DIP (TOP VIEW)

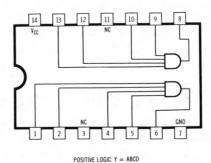

POSITIVE LOGIC: Y = ABCD

(B) Type 9H21.

DIP (TOP VIEW)

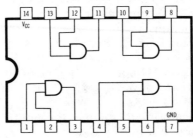

POSITIVE LOGIC: Y = AB

(C) Type 9N09.

Fig. 1-13. Examples of AND gates. (*Courtesy Fairchild Camera and Instrument Corp.*)

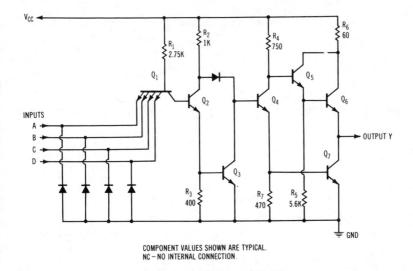

COMPONENT VALUES SHOWN ARE TYPICAL.
NC – NO INTERNAL CONNECTION.

Fig. 1-14. Basic AND-gate schematic diagram. (*Courtesy Fairchild Camera and Instrument Corp.*)

We will find as we proceed that some digital logic diagrams employ another definition of logic states called *negative logic*. We will also find that TTL is not the only family of logic chips (ICs). For our present purposes, it is desirable to confine our descriptions to positive logic, wherein a comparatively high positive voltage represents the logic 1 state and a comparatively low positive voltage (or ground potential) represents the 0 state.

Note in passing that most AND-gate packages contain gates with the same number of inputs. However, the technician will occasionally encounter a package with a two-input AND gate and a three-input AND gate, for example.

AND GATE COMBINATIONS

Various AND gate combinations are shown in Fig. 1-15. For example, a pair of two-input AND gates may be connected to function as a three-input AND gate (Fig. 1-15A). Similarly, three two-input AND gates may be connected to function as a four-input AND gate (Fig. 1-15B). Observe that unused inputs (Figs. 1-15C, D, and E) may be tied to a used input or to V_{CC} to stabilize circuit operation (remember the problem of a floating input). There is one practical point that is worth keeping in mind: not all TTL gates are designed to withstand the full V_{CC} supply voltage on the inputs. Therefore, in such a case, an un-

21

used input is tied to a logic-high point in the system instead of to V_{CC}.

"STUCK AT"

A practical troubleshooting note concerning a "stuck at" trouble symptom is illustrated in Fig. 1-16. A bond is an internal

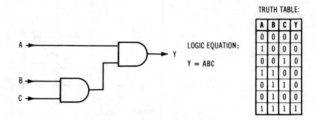

LOGIC EQUATION:

Y = ABC

TRUTH TABLE:

A	B	C	Y
0	0	0	0
1	0	0	0
0	0	1	0
1	1	0	0
0	1	1	0
0	1	0	0
1	1	1	1

(A) Equivalent of three-input AND gate.

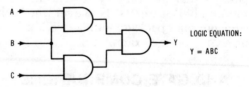

LOGIC EQUATION:

Y = ABCD

(B) Equivalent of four-input AND gate.

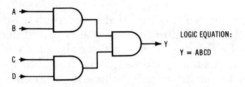

LOGIC EQUATION:

Y = ABC

(C) Equivalent of three-input AND gate.

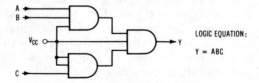

LOGIC EQUATION:

Y = ABC

(D) Equivalent of three-input AND gate.

Fig. 1-15. Some examples

connection in an IC package. If a bond open-circuits, the point in the IC to which it was connected is not externally accessible. In the example under consideration, the bond is open-circuited between points A and B. Accordingly, the gate "sees" point B as a logic-high level, and input terminal A is said to be "stuck at" a

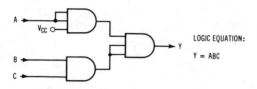

(E) Equivalent of three-input AND gate.

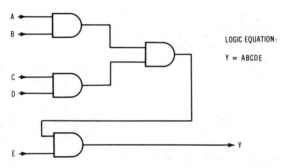

(F) Equivalent of five-input AND gate.

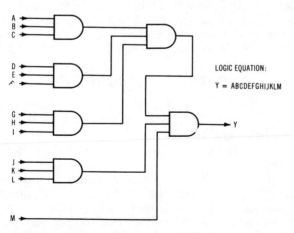

(G) Equivalent of thirteen-input AND gate.

of combinations of AND gates.

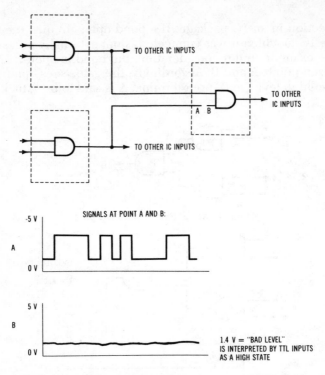

Fig. 1-16. An AND gate with an open input bond. (*Courtesy Hewlett-Packard*)

high level. In other words, if the A input terminal is driven logic-low and the other gate input is driven logic-high, the output does not go low as in normal operation. Other examples of "stuck at" trouble symptoms are discussed in later chapters.

AND GATE APPLICATIONS

An application for an AND gate in a video cassette recorder (vcr) is shown in Fig. 1-17. Signals derived from the tape-end sensing oscillator and the drum pulse generator are applied to an AND gate. If both inputs to the gate go high, the stop-solenoid driver is energized.

Another typical application for an AND gate is shown in Fig. 1-18. The AND gate operates in a video game and functions to form a pulse for display of a spot on the picture-tube screen when coincident horizontal and vertical pulses are applied to the gate. The horizontal pulse has a screen position left-to-right that is determined by the value of the horizontal-position con-

24

trol voltage. The vertical pulse has a screen position up-and-down that is determined by the value of the vertical-position control voltage. If the AND gate were not used, the pulses would display bars on the screen; however, the gate functions to produce zero output except for the short time during which the horizontal and vertical "bars" are coincident.

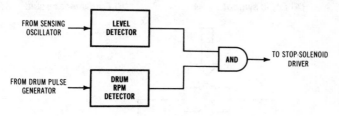

Fig. 1-17. Application of AND gate in vcr.

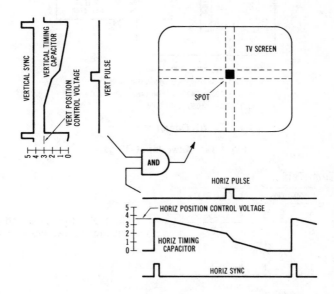

Fig. 1-18. Use of AND gate to produce spot display.

OR GATES

Another basic type of gate performs the OR function. Operation of a two-input OR gate is shown in Fig. 1-19. Observe that the output of the OR gate goes logic-high if either or both of its inputs are driven logic-high. This function is set forth in the truth table and the logic equation (Fig. 1-20). Note that the logic

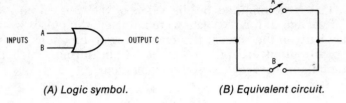

INPUTS A B → OUTPUT C

(A) Logic symbol.

(B) Equivalent circuit.

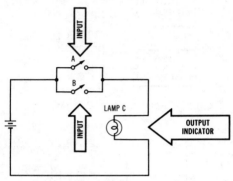

1. BOTH SWITCHES OPEN: LAMP IS DARK.
2. SWITCH A CLOSED, SWITCH B OPEN: LAMP GLOWS.
3. SWITCH B CLOSED, SWITCH A OPEN: LAMP GLOWS.
4. BOTH SWITCHES CLOSED: LAMP GLOWS.

(C) Operation.

Fig. 1-19. A two-input OR gate.

Truth Table:

INPUTS		OUTPUT
A	B	C
0	0	0
0	1	1
1	0	1
1	1	1

Fig. 1-20. Truth table and logic equation for two-input OR gate.

Logic Equation:

$$C = A + B$$

equation is read, "C equals A OR B." (This is a Boolean-algebra equation in which the "plus" sign denotes the OR function. Recall that in the Boolean-algebra equation for an AND gate, the "multiplication" sign [·] denotes the AND function.) Although it seems a bit confusing at first to call a "plus" sign OR and to call a "times" sign AND, we will soon be taking these new conventions as a matter of course.

26

AND-OR GATE COMBINATIONS

Combinations of AND and OR gates are widely used in digital-logic circuits. Basic arrangements are depicted in Fig. 1-21. Observe that an OR gate can be compared with a mixer; in other words, an OR gate produces an output that "follows" each of its inputs. The OR gate also provides isolation between its inputs; that is, if the OR gate were omitted in the first example of Fig. 1-21, AND gates U1 and U2 would not operate normally. If the outputs from U1 and U2 were directly connected together, the

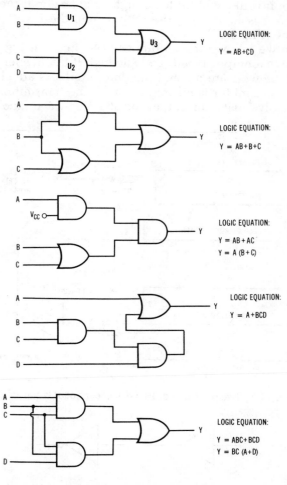

Fig. 1-21. Combinations of AND and OR gates.

output from U2 could not be driven high if the output from U1 happened to be low. Conversely, the output from U1 could not be driven high if the output from U2 happened to be low. Thus, U3 provides isolation and prevents interaction between U1 and U2. As in previous examples, the logic equations in Fig. 1-21 represent positive logic.

An example of the application of AND and OR gates in television-transmitter operation is shown in Fig. 1-22. This circuit includes an inverter; note that an inverter is the same kind of device as a buffer, except that an inverter reverses the input signal. That is, if the input to an inverter is logic-low, the output from the inverter will be logic-high; conversely, if the input to an inverter is logic-high, the output from the inverter will be logic-low (Fig. 1-23).

When we follow through the AND-OR circuit in Fig. 1-22, we observe that output H will go logic-high *if* the filaments are on, and the blowers are on, and the door interlocks are closed, *unless* the transmitter is off frequency, or the transmitter is drawing excessive plate current, or the standing-wave ratio is exces-

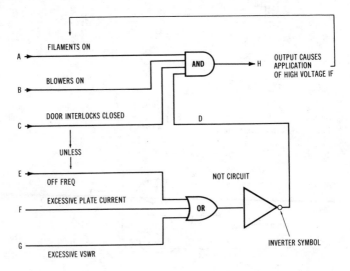

Fig. 1-22.   Use of AND and OR gates in control of tv transmitter.

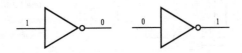

Fig. 1-23.   Reversal of logic states by inverter.

28

sive. If any of the last three conditions exists, the application of high voltage will be inhibited.

LOGIC FLOW DIAGRAM

Consider next the logic flow diagram (Fig. 1-24) for the AND-OR gate application described above. Observe that a logic flow diagram shows the logic states at every node (interconnection point) from input to output of the circuit. These states are expressed as logic equations. In particular, take note of the NOT notation in the equations. The NOT function is indicated by a line, called a *bar*, over the term or terms that are affected. For example, an inverter is a NOT operator; if the input to an inverter is written A, then the output from the inverter may be written $\overline{A}$. The term $\overline{A}$ is read, "NOT A." If the input to an inverter is written $\overline{A}$, then the output from the inverter is written A.

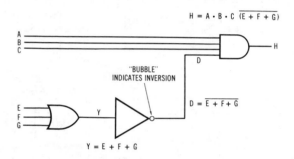

Fig. 1-24. Logic flow diagram for the example of Fig. 1-22.

With respect to the logic flow diagram in Fig. 1-24, the input to the inverter is written $E+F+G$, and the output from the inverter is written $\overline{E+F+G}$. The expression $\overline{E+F+G}$ is read, "NOT E OR F OR G." Finally, the output from the AND gate is written $A \cdot B \cdot C\overline{(E+F+G)}$; alternatively, this expression may be written $ABC\overline{(E+F+G)}$. The expression is read, "A AND B AND C AND NOT E OR F OR G." As detailed later, it would be incorrect to read $\overline{E+F+G}$ as "NOT E OR NOT F OR NOT G," because the logical meaning would be changed.

LEVEL INDICATORS

At this point, it is helpful to take note of level indicators used in logic diagrams. We have seen that if a buffer is followed by a

29

small circle ("bubble") a logic-high input level will be changed into a logic-low output level, or vice versa. Accordingly, the small circle is a logic-level indicator. We say that the bubble *negates* the logic level (reverses the logic level). We will find that the *input* to a logic device, such as a gate or a buffer, may be negated. Let us review the logical meanings of negated inputs and negated outputs.

A level indicator in a digital-logic configuration serves to indicate the logic level that will exist for the *intended function of the device*. For example, an inverter may be shown as a buffer with negated input, or it may be shown as a buffer with negated output (Fig. 1-25A). In a general sense, these two symbols are the same—a given input logic level corresponds to an opposite output logic level. In a more specialized sense, however, the two symbols serve to distinguish between a logic-high input level and a logic-low input level *for the intended output level*.

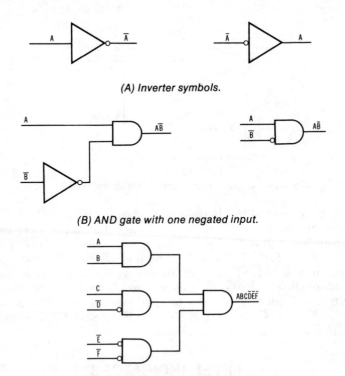

(A) Inverter symbols.

(B) AND gate with one negated input.

(C) AND gates with some negated inputs.

Fig. 1-25. Indication of logic levels.

30

Let us take a closer look at circuit activity and the correspond-ing level indication. If an inverter is drawn as a buffer with a bubbled output, this means that the following logic device or circuit will be triggered (or will be otherwise active) when a logic-high state is applied to the input of the inverter. On the other hand, if an inverter is drawn as a buffer with negated (bubbled) input, this means that the following logic device or circuit will be active when a logic-low state is applied to the input of the inverter. In the first case, we say that the circuit is active-high; in the second case, we say that the circuit is active-low.

Consider next an AND gate with one input negated (Fig. 1-25B). It is evident that the circuit will be active only when the A input is logic-high and the B input is logic-low. Therefore, the level indicators for the AND gate are written as A, $\overline{B}$, and $A\overline{B}$. The basic importance of appropriate logic-level indication is apparent in Fig. 1-25C, which shows an AND gate, an AND gate with one negated input, and an AND gate with both inputs ne-gated. There are 64 possible state combinations of the six inputs to this circuit; however, only one of these combinations will produce a logic-high output. This combination is $ABC\overline{D}\overline{E}\overline{F}$, which is read, "A AND B AND C AND NOT D AND NOT E AND NOT F."

It is clear that digital logic diagrams become much easier to read when level indicators are chosen to indicate the logic levels that will exist for the intended function of a device or circuit.

Chapter
2

Additional Logic Gates and Circuit Operation

The OR gate function was noted in the first chapter. Like AND gates, OR gates are packaged as two-input devices, three-input devices, or occasionally as two-input and three-input devices. A typical package is represented in Fig. 2-1. This is an example of the widely used TTL logic family, and the logic equation is stated for positive logic. Prior mention was made of TTL logic-high and logic-low levels; these are fundamental considerations in digital-logic troubleshooting procedures. Therefore, let us observe a typical digital waveform that would be displayed at the input of an OR gate (or any other type of gate, for that matter).

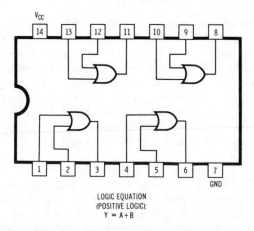

LOGIC EQUATION
(POSITIVE LOGIC):
Y = A + B

Fig. 2-1.   Pinout diagram of Type 9N32 quad two-input OR gate.

33

INPUT AND OUTPUT WAVEFORMS

Fig. 2-2 shows an example of a TTL digital waveform, with the standard logic-high (1) and logic-low (0) levels indicated. Note that a logic-high level is at least +2.4 volts and that a logic-low level is not greater than +0.4 volt. Thus, +0.6 volt would be a "bad level," as would +1 volt or +1.5 volts. As indicated in the diagram, a typical digital waveform shows various departures from the ideal. The top and bottom excursions are not perfectly flat, the leading edges do not rise instantaneously, and the trailing edges do not fall instantaneously. Overshoot and ringing are often seen in a digital waveform. Not shown in Fig. 2-2 is another trouble condition, excessive noise contamination. This consideration is described in more detail in a later chapter.

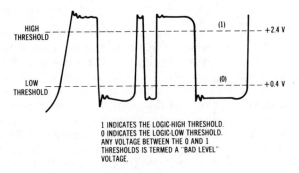

1 INDICATES THE LOGIC-HIGH THRESHOLD.
0 INDICATES THE LOGIC-LOW THRESHOLD.
ANY VOLTAGE BETWEEN THE 0 AND 1
THRESHOLDS IS TERMED A "BAD LEVEL"
VOLTAGE.

Fig. 2-2. A typical TTL input or output digital waveform.

Digital waveforms are displayed on the screen of an oscilloscope (often simply called a "scope"), an example of which is shown in Fig. 2-3. It is desirable to employ triggered sweep and to have a calibrated time base available so that pulse widths and rise times can be measured (when these parameters are too far out of normal tolerance, system malfunction results). A bandwidth of 10 MHz is adequate for most digital troubleshooting tasks. It is helpful to have dual-trace operation available; this facilitates comparative pulse-timing tests within a digital system.

A typical application of OR gates in a graphic terminal is shown in Fig. 2-4. (A graphic terminal functions to drive a crt display, as in a tv typewriter for a personal computer.) An oscilloscope is very helpful in troubleshooting this kind of circuitry. The gates serve as combiners, or mixers, for the sync pulses and

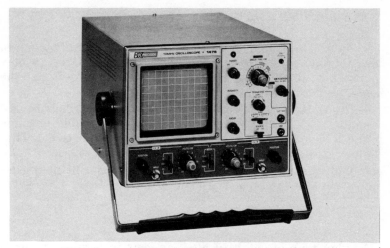

Fig. 2-3.   A dual-trace, 10-Mhz triggered oscilloscope.
(*Courtesy B & K Precision*)

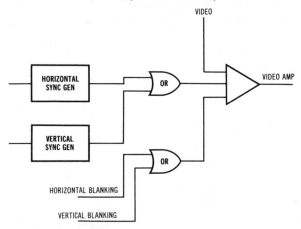

Fig. 2-4.   An OR-gate application in a tv typewriter.

blanking pulses but prevent the pulse generators from interacting with each other. Observe that the upper OR gate serves to isolate the horizontal-sync generator from the vertical-sync generator. The lower OR gate serves to isolate the horizontal-blanking section from the vertical-blanking section. If vertical sync is lost, a scope will quickly show whether the trouble is in the vertical-sync generator or in the upper OR gate. Similarly, loss of horizontal sync or loss of blanking can be quickly tracked down to a source or to a gate.

THREE BASIC OPERATIONS

The three basic logic operations are called the *conjunction* or logical product (AND), symbolized as in AB or A·B; the *disjunction* or logical sum (OR), symbolized by +, as in A+B; and the *negation* (NOT), symbolized by bar notation, as in $\overline{AB}$. Thus, $\overline{1}=0$, $\overline{0}=1$, $1001=\overline{0110}$, and $\overline{A}\overline{B}\overline{C}=\overline{A}\overline{\overline{B}}\overline{C}$. Observe that the last example is an operation of *double negation*. Double negation brings a logic term or expression back to where it started. Thus $\overline{\overline{B}}=B$. Note than an inverter is not a gate; an inverter is an operator. Only the AND gate and the OR gate are basic; all other gates are essentially combinatorial logic configurations.

XOR GATES

Another type of gate that often appears in logic diagrams is termed the exclusive OR (XOR) gate (Fig. 2-5). An XOR gate always has two, and only two, inputs. Operation of an XOR gate is shown in Fig. 2-6. An XOR gate differs from an OR gate in that it

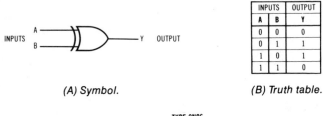

INPUTS		OUTPUT
A	B	Y
0	0	0
0	1	1
1	0	1
1	1	0

(A) Symbol.   *(B) Truth table.*

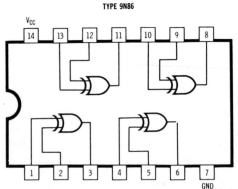

(C) Pinout diagram of quad XOR gate.

Fig. 2-5. Exclusive OR (XOR) gate.

provides a logic-high output only when *opposite* logic levels are applied to its inputs. If both inputs are logic-low, or if both inputs are logic-high, the output of the XOR gate is logic-low.

The XOR function is clarified by the series-parallel switching circuit (basic equivalent circuit) depicted in Fig. 2-6. The logic equation for an XOR gate may be written $Y = A\overline{B} + \overline{A}B$ and is read, "Y equals A AND NOT B OR NOT A AND B." Alternatively, the logic equation for the XOR gate may be written $Y = A \oplus B$, which is read, "Y equals A exclusive-OR B."

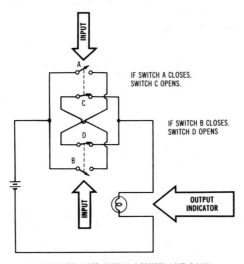

1. IF NEITHER SWITCH A NOR SWITCH B IS PRESSED, LAMP IS DARK.
2. IF ONLY SWITCH A IS PRESSED, LAMP GLOWS.
3. IF ONLY SWITCH B IS PRESSED, LAMP GLOWS.
4. IF BOTH SWITCH A AND SWITCH B ARE PRESSED, LAMP IS DARK.

LOGIC EQUATION:
$$Y = A \oplus B = A\overline{B} + \overline{A}B$$

Fig. 2-6. Operation of an XOR gate.

Insofar as digital logic is concerned, the XOR function is not regarded as a basic function, but as a combinatorial logic function. In other words, the XOR function is a combination of AND and OR functions. Observe the implementations for the XOR function depicted in Fig. 2-7. Here, negation is performed by inverters; either one or two inverters may be utilized in combination with two AND gates and an OR gate. Note that both configurations perform the XOR function and that $B\overline{C} + \overline{B}C$ makes the same basic statement as $(A+B)\overline{AB}$. The latter logic equation is read, "A OR B AND NOT A AND B."

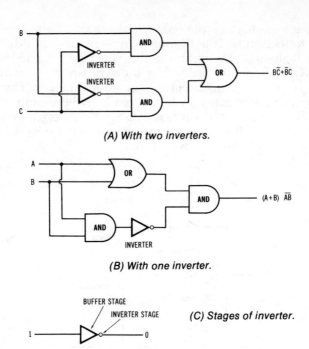

(A) With two inverters.

(B) With one inverter.

BUFFER STAGE
INVERTER STAGE *(C) Stages of inverter.*

Fig. 2-7. Examples of combinatorial logic to perform XOR function.

Exclusive OR gates are used, for example, in digital test instruments as pulse phase detectors. One type of in-circuit transistor tester employs XOR-gate circuitry to indicate if the collector output from a transistor is 180° out of phase with the base input pulse waveform, if the input and output waveforms are in phase, or if they have a phase difference in the range from 0° to 180°. The basic waveform relations are shown in Fig. 2-8. If the inputs to an XOR gate are 180° out of phase, the output is logic-high (Fig. 2-8B). If the inputs are in phase, the output is logic-low (Fig. 2-8C). Intermediate phase differences result in outputs with different duty cycles (Figs. 2-8D and E); the average values of these waveforms differ, and a meter responds with a higher reading to a waveform with a higher duty cycle.

The logic circuitry shown in Fig. 2-9 is the heart of the Sencore TF26 transistor tester. The transistor is driven by a 2-kHz square wave, and its response is evaluated by the logic circuitry. Base and collector signals from the transistor under test are amplified and then applied to NAND gate IC2A, which is part of a quad two-input NAND gate wired as an XOR gate. A NAND gate has a low output only when both inputs are high. The base signal from the transistor under test is coupled into input pin 2 of IC2A; if the transistor is good, the base signal will be 180° out of phase with the collector signal coupled into input pin 1. Accordingly, both inputs of IC2A do not go high simultaneously, and the output from IC2A remains high. This high output is coupled into input pin 9 of IC2B and input pin 5 of IC2C. These gates function as inverters and couple the inverted base and collector signals into the inputs of IC2D. Accordingly, both inputs of IC2D do not go high simultaneously, and the output from IC2D remains high. A "good" indication is thereby obtained. On the other hand, when a bad transistor is under test, the inputs of IC2A will be in phase, resulting in a square-wave output from IC2A that is 180° out of phase with the input. The net result is a low output from IC2D.

NAND GATES

Another kind of gate that is often encountered in digital logic diagrams is the NAND gate (Fig. 2-10). A NAND gate is an AND gate followed by an inverter (Fig. 2-11). Therefore, the truth table and logic equation for a NAND gate are similar to those for an AND gate except that the output terms are inverted. It is often said that an inverter *complements* the input logic state. Similarly, a NAND gate provides the *complement* of an AND-gate output. The complement of 1 is 0; the complement of 0 is 1; the complement of 1 is $\overline{1}$; the complement 0 is $\overline{0}$. Common NAND-gate IC package pinouts are shown in Fig. 2-12.

An example of an application for NAND gates in a home appliance is shown in Fig. 2-13. This is a fireplace monitor that is actuated by a phototransistor; LED output indication is provided. A simple arrangement of this kind can easily be checked out by means of *static tests*. If the input of IC2C is logic-high, then the inverter output will normally be logic-low; if both inputs of IC1A are logic-high, then the NAND-gate output will normally be logic-low. A tvm or dvm serves as a good indicator of static logic states.

A static test is so-called because only dc voltages are present throughout the circuit. By way of comparison, most digital-logic circuits are characterized by both dc voltages and ac (pulse) voltages. In turn, troubleshooting of most digital logic circuits involves both static tests and dynamic tests. These procedures are detailed in later chapters.

Previous mention has been made of "stuck-at" trouble symptoms. Both *external* and *internal* fault conditions can cause these symptoms. For example, if the output of a NAND gate is

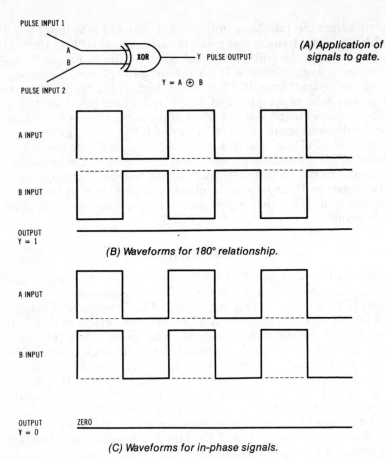

PULSE INPUT 1

A
B

XOR — Y PULSE OUTPUT

(A) Application of signals to gate.

Y = A ⊕ B

PULSE INPUT 2

A INPUT

B INPUT

OUTPUT
Y = 1

(B) Waveforms for 180° relationship.

A INPUT

B INPUT

OUTPUT
Y = 0

ZERO

(C) Waveforms for in-phase signals.

Fig. 2-8.   Operation of XOR

"stuck high" (or "stuck low") and does not respond to the input logic states, the trouble may be found in the printed-circuit wiring (external fault), or it may be tracked down to a defective NAND gate (internal fault). As an illustration, if one input conductor of a NAND gate is shorted to ground, the output of the gate will be "stuck high." Also, if the output conductor of a NAND gate is shorted to V_{CC}, the output of the gate will be "stuck high."

Internal faults in gates include open bonds, as previously noted, and also steering-circuit defects. Fig. 2-14 shows as an example a TTL NAND gate that employs four transistors and a diode. In this circuit, Q2 is called the *steering transistor*. It drives output transistors Q3 and Q4. If Q2 happens to develop a

40

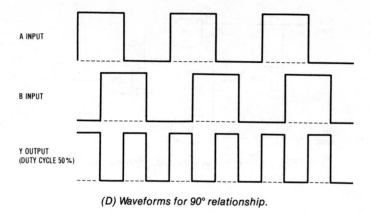

(D) Waveforms for 90° relationship.

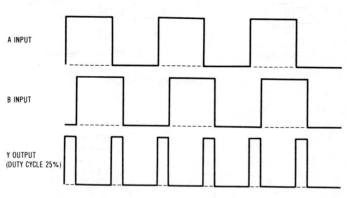

(E) Waveforms for 45° relationship.

gate in phase detector.

collector-base short, Q4 will remain in saturation and Q3 will remain cut off, regardless of the input logic states; the gate output will be "stuck low." Replacement of the NAND gate is required, inasmuch as the steering circuitry is inaccessible.

NEGATED AND GATES

A distinction must be made between a NAND gate and a negated AND gate. Refer to Fig. 2-15. A NAND gate consists of an AND gate followed by an inverter, whereas a negated AND gate consists of a pair of inverters followed by an AND gate. Observe in particular that the truth tables for these two types of gates are not the same. The logic equation for the NAND gate is written

41

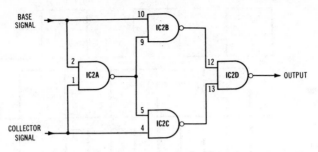

Fig. 2-9. Logic circuitry in transistor tester.

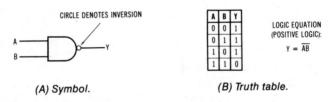

CIRCLE DENOTES INVERSION

A	B	Y
0	0	1
0	1	1
1	0	1
1	1	0

LOGIC EQUATION
(POSITIVE LOGIC):

$Y = \overline{AB}$

(A) Symbol.   *(B) Truth table.*

Fig. 2-10. Two-input NAND gate.

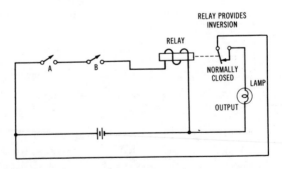

Fig. 2-11. An equivalent circuit for a NAND gate.

$Y=\overline{AB}$, whereas the logic equation for the negated AND gate is written $Y=\overline{A}\cdot\overline{B}$, or $Y=\overline{A}\overline{B}$. The logic equation for the NAND gate is read, "Y equals NOT A AND B," whereas the logic equation for the negated AND gate is read, "Y equals NOT A AND NOT B." Both types of gates are often encountered in logic diagrams.

A negated AND gate may have all of its inputs negated, or it may have only one of its inputs negated. A negated AND gate with four inputs may have two or three of its inputs negated. A practical example of a two-input negated AND gate with only one of its inputs negated is shown in Fig. 2-16. When the rewind switch is closed, input A goes logic-high, thereby "arming," "enabling," or "cocking" the gate. While the tape is rewinding, the phototransistor is not illuminated, and input B is logic-high. However, when the end of the tape passes the phototransistor and permits light to enter, input B goes logic-low. In turn, the gate output goes logic-high and actuates the stop solenoid.

An electromechanical equivalent circuit for a two-input AND gate with one negated input is shown in Fig. 2-17. The relay operates as an inverter and negates the B switch. If both inputs of the gate were negated, the A switch would also be connected to a relay winding. In turn, the lamp would glow when both switches were open; however, if one or both of the switches were closed, the lamp would be dark.

NOR GATES

A NOR gate consists of an OR gate followed by an inverter (Fig. 2-18). The truth table for a NOR gate is the same as the truth table for an OR gate, except that the output logic states are inverted. Similarly, the logic equation for a NOR gate is the same as the logic equation for an OR gate, except that the output term is negated, or inverted. For example, the logic equation for a three-input NOR gate is written $Y=\overline{A+B+C}$; this equation is read, "Y equals NOT A OR B OR C." An equivalent circuit for a NOR gate is shown in Fig. 2-19.

Now, refer to Fig. 2-15B and observe that the truth table for a negated AND gate is exactly the same as the truth table for a NOR gate (Fig. 2-18B). It is important to recognize that a negated AND gate performs the NOR function. Accordingly, the three negated AND gates in Fig. 2-20 should be considered to be NOR gates when they appear in a positive-logic diagram. (All of the diagrams discussed so far have been examples of positive logic.)

NEGATED OR FUNCTION

Note the comparison of NOR and negated OR gates shown in Fig. 2-21. A NOR gate consists of an OR gate followed by an inverter, whereas a negated OR gate consists of a pair of inverters followed by an OR gate. Note that the truth tables for these two types of gates are not the same. The logic equation for the NOR gate is written $Y=\overline{A+B}$, whereas the logic equation for the ne-

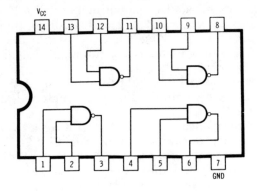

TYPE 9N100

POSITIVE LOGIC: Y = $\overline{AB}$

(A) Quad two-input NAND gate.

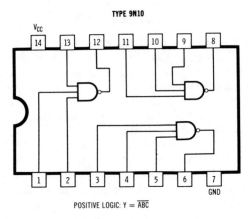

TYPE 9N10

POSITIVE LOGIC: Y = $\overline{ABC}$

(B) Triple three-input NAND gate.

Fig. 2-12. Common NAND-

gated OR gate is written $Y=\overline{A}+\overline{B}$. The logic equation for the NOR gate is read, "Y equals NOT A OR B," whereas the logic equation for the negated OR gate is read, "Y equals NOT A OR NOT B." Now, refer back to the truth table for the NAND gate in Fig. 2-10, and observe that the truth tables for the NAND gate and the negated OR gate are the same. This leads to an important law of logic: *A negated OR gate performs the NAND function.* Accordingly, the gates depicted in Fig. 2-22 are all NAND gates with negated OR implementation.

44

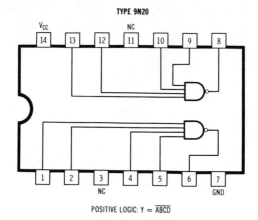

POSITIVE LOGIC: Y = $\overline{ABCD}$

(C) Dual four-input NAND gate.

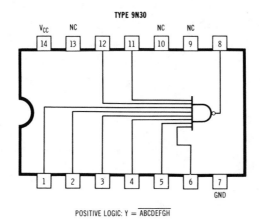

POSITIVE LOGIC: Y = $\overline{ABCDEFGH}$

(D) Eight-input NAND gate.

gate package pinouts.

DEMORGAN'S LAWS FOR EQUIVALENT GATES

All equivalent gates obey DeMorgan's laws. These laws can be quickly verified by comparison of the truth tables for the various devices. Thus:

1. A negated NAND gate performs the OR-gate function.
2. A negated NOR gate performs the AND-gate function.
3. A negated OR gate performs the NAND-gate function.

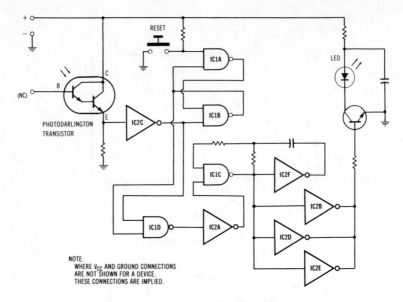

Fig. 2-13.   Circuit of a fireplace monitor.

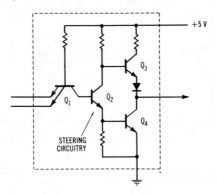

Fig. 2-14.   Steering circuit in a TTL NAND gate.

4. A negated AND gate performs the NOR-gate function.

A helpful summary of DeMorgan's laws states: An AND gate can be substituted for an OR gate, provided that all of the logic-level indicators are reversed (Figs. 2-23A and B). Similarly, an OR gate may be substituted for an AND gate, provided that all of the logic-level indicators are reversed (Figs. 2-23C and D). Fig. 2-24 provides a summary of equivalent gates.

46

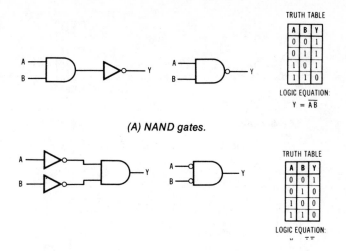

TRUTH TABLE

A	B	Y
0	0	1
0	1	1
1	0	1
1	1	0

LOGIC EQUATION:

$Y = \overline{AB}$

(A) NAND gates.

TRUTH TABLE

A	B	Y
0	0	1
0	1	0
1	0	0
1	1	0

LOGIC EQUATION:

(B) Negated AND gates.

Fig. 2-15. Comparison of NAND gates and negated AND gates.

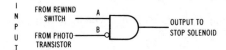

Fig. 2-16. Control circuit for a video cassette recorder.

(A) Symbol and logic equation.

LOGIC EQUATION
(POSITIVE LOGIC)

$Y = A\overline{B}$

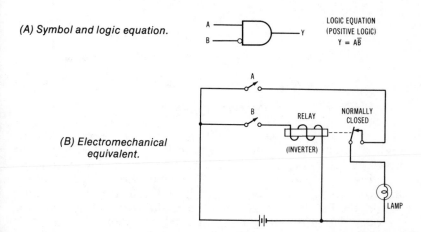

(B) Electromechanical
equivalent.

Fig. 2-17. An electromechanical equivalent circuit for a two-input AND
gate with one negated input.

47

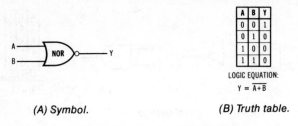

A	B	Y
0	0	1
0	1	0
1	0	0
1	1	0

LOGIC EQUATION:

$Y = \overline{A + B}$

(A) Symbol.   *(B) Truth table.*

Fig. 2-18.   Two-input NOR gate.

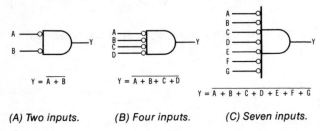

Fig. 2-19.   Equivalent circuit for a NOR gate.

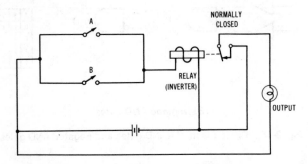

$Y = \overline{A + B}$ $Y = \overline{A + B + C + D}$

$Y = \overline{A + B + C + D + E + F + G}$

(A) Two inputs.   *(B) Four inputs.*   *(C) Seven inputs.*

Fig. 2-20.   Examples of NOR gates with negated AND implementation.

XNOR GATES

Exclusive NOR (XNOR or $\overline{XOR}$) gates are also encountered in
logic diagrams. An XNOR gate is basically an XOR gate followed
by an inverter (Fig. 2-25). The logic equation for an XNOR gate
is written $Y = \overline{A \oplus B}$ and is read, "Y equals NOT A exclusive-OR
B." Note in passing that if the inputs to an XOR gate are inverted, the function remains the same. Similarly, if the inputs to
an XNOR gate are inverted, the function remains the same.

Fig. 2-26 shows a logic diagram of an IC package that contains
four XOR gates. Notice that two of the gates provide XOR outputs
and the other two gates provide both XOR and XNOR ($\overline{XOR}$) out-

48

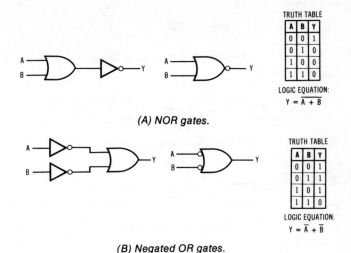

TRUTH TABLE

A	B	Y
0	0	1
0	1	0
1	0	0
1	1	0

LOGIC EQUATION:
$Y = \overline{A + B}$

(A) NOR gates.

TRUTH TABLE

A	B	Y
0	0	1
0	1	1
1	0	1
1	1	0

LOGIC EQUATION:
$Y = \overline{A} + \overline{B}$

(B) Negated OR gates.

Fig. 2-21.   Comparison of NOR gates and negated OR gates.

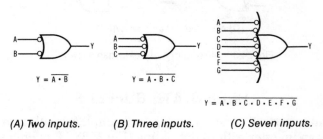

$Y = \overline{A \cdot B}$ $Y = \overline{A \cdot B \cdot C}$

$Y = \overline{A \cdot B \cdot C \cdot D \cdot E \cdot F \cdot G}$

(A) Two inputs.   *(B) Three inputs.*   *(C) Seven inputs.*

Fig. 2-22.   Examples of NAND gates with negated OR implementation.

puts. This is an example of an exception to the general rule that a gate has only one output.

AND-OR-INVERT GATES

A widely used form of packaged combinatorial logic called the AND-OR-INVERT gate is shown in Fig. 2-27. It consists of AND gates followed by a NOR gate. As seen from the logic equation, an AND-OR-INVERT gate is equivalent to a NOR gate with several ANDed inputs. The arrangement is *not* termed an "AND-NOR" gate, as might be assumed from its circuitry. Four AND gates are utilized in this example; three of the AND gates have two inputs, and one of the AND gates has three inputs. This arrangement is called a four-wide 2-2-2-3–input AND-OR-INVERT gate.

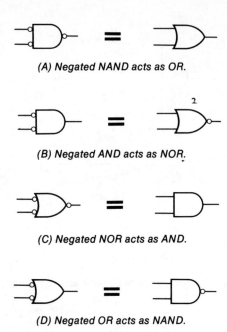

(A) Negated NAND acts as OR.

(B) Negated AND acts as NOR.

(C) Negated NOR acts as AND.

(D) Negated OR acts as NAND.

Fig. 2-23.   Illustrations of DeMorgan's laws.

THREE-STATE BUFFERS

A three-state (Tri-State®) buffer is an amplifier with a control line or an inverter with a control line (Fig. 2-28). It operates like a conventional buffer or inverter as long as the control input is held logic-high (Figs. 2-29A and B); when the control input is driven logic-low, the output terminal of the device is effectively disconnected from its load circuit (Fig. 2-29C).

Three-state buffers are widely used in bus-organized computer systems. The bus can carry digital signals in either direction, but only in one direction at a given time. Also, the bus is connected to a number of digital signal sources. The system can be compared to a party line that is time-shared by a number of telephone subscribers. Only one pair of talkers and listeners uses the party line at any given time; the other subscribers hang up the receiver and switch their phones off the party line. Similarly, a three-state buffer serves to switch a digital signal source into or out of a bus. When the control input of a three-state buffer is driven logic-low, the "receiver is on the hook"; when the control input is driven logic-high, the "receiver is off the hook."

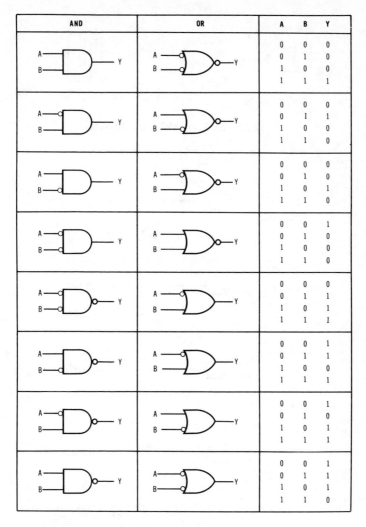

Fig. 2-24. Summary of equivalent gates.

SERIES-PARALLEL INVERTER CIRCUITRY

Buffers or inverters may be connected in parallel to provide more current output capability. They may be connected in series to obtain a desired logic-state relation. These devices are sometimes configured in series-parallel, as in the example of Fig. 2-30. This is a circuit arrangement for a simple digital-logic probe for use in TTL circuits. This probe is somewhat more

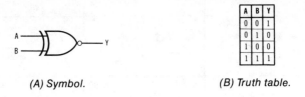

A	B	Y
0	0	1
0	1	0
1	0	0
1	1	1

(A) Symbol.   (B) Truth table.

Fig. 2-25. Exclusive NOR (XNOR) gate.

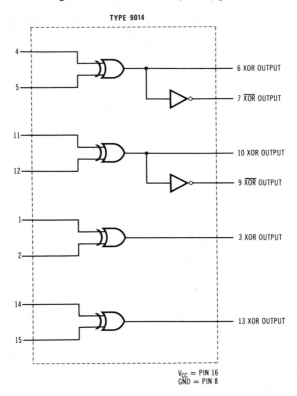

Fig. 2-26. Example of XOR and XNOR IC. (*Courtesy Fairchild Camera and Instrument Corp.*)

elaborate than the earlier example, because two LEDs are employed. Each LED is driven by a pair of inverters to supply the current demand. Neither LED will glow when the input is open-circuited ("floating"). One LED glows if the input is logic-low, and the other LED glows if the input is logic-high. A CD 4009 COSMOS hex inverter IC package may be used with MV5020 LEDs.

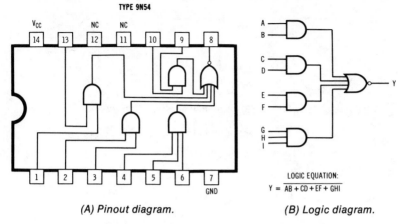

(A) Pinout diagram.

LOGIC EQUATION:
$$Y = \overline{AB + CD + EF + GHI}$$

(B) Logic diagram.

Fig. 2-27. Example of AND-OR-INVERT gate.

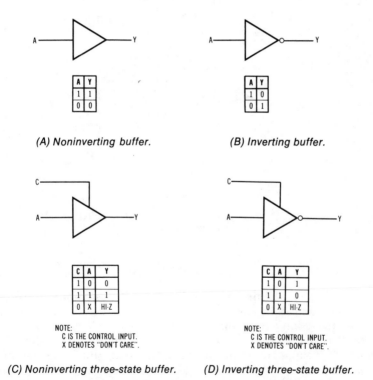

(A) Noninverting buffer.

A	Y
1	1
0	0

(B) Inverting buffer.

A	Y
1	0
0	1

C	A	Y
1	0	0
1	1	1
0	X	HI-Z

NOTE:
C IS THE CONTROL INPUT.
X DENOTES "DON'T CARE".

(C) Noninverting three-state buffer.

C	A	Y
1	0	1
1	1	0
0	X	HI-Z

NOTE:
C IS THE CONTROL INPUT.
X DENOTES "DON'T CARE".

(D) Inverting three-state buffer.

Fig. 2-28. Conventional and three-state buffers.

(A) Logic-high output.   (B) Logic-low output.   (C) High-impedance output.

Fig. 2-29.   Equivalent output circuit of a three-state buffer.

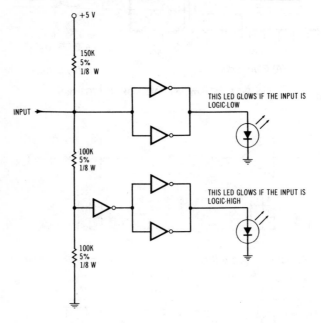

Fig. 2-30.   Example of series-parallel inverting buffer circuitry
used in digital logic probe.

DISCRETE LOGIC

The digital technician may occasionally encounter *discrete circuits,* such as the one shown in Fig. 2-31A. A discrete circuit is one in which each element is separately packaged. A discrete circuit is many times larger than an integrated circuit that performs the same function, but it is tested in the same manner. Note that the operating voltages in a discrete circuit may be somewhat different from those in an equivalent integrated circuit.

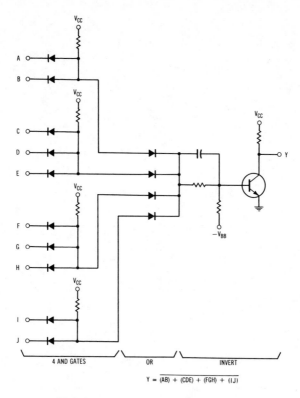

$$Y = \overline{(AB) + (CDE) + (FGH) + (IJ)}$$

(A) Schematic diagram of discrete circuit.

(B) Logic diagram of integrated circuit.

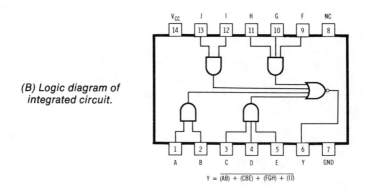

$$Y = \overline{(AB) + (CBE) + (FGH) + (IJ)}$$

Fig. 2-31.   Two versions of an AND-OR-INVERT gate. (*Courtesy Hewlett-Packard*)

55

Chapter
3

Basic Adders and Subtracters

Two basic mathematical operations are addition and subtraction. This chapter deals with the configurations of logical elements that are used to carry out these operations in digital circuits.

TIMING DIAGRAMS

When gates are interconnected to perform comparatively involved functions, a *timing diagram* is often an important adjunct to a *logic diagram*. A logic diagram represents the logical elements and their interconnections in a digital system. It is a pictorial representation of interconnected logic elements using standard symbols to represent logic functions. On the other hand, a timing diagram, also called a *timing chart*, shows the waveform relations for a digital device, circuit, or system. The timing diagram sets forth precisely how logical elements and their configurations operate. In Fig. 3-1, representative input/output waveforms are shown for AND, NAND, OR, and NOR gates. These waveforms provide the basis for standard timing diagrams, such as the one shown in Fig. 3-2. Waveforms are presented in their relative time relations, one beneath another.

Consider next the timing diagram for an XOR gate, shown in Fig. 3-3. The gate output is logic-low when both inputs are logic-high or logic-low. The gate output is logic-high if one input is logic-high and the other input is logic-low. Note that if the A input is held logic-high constantly while the B input is rapidly pulsed high-low-high-low . . . , the gate output will "follow" the B input with a complementary pulse train low-high-low-high . . . as long as the B input is driven. In other words, a digital system is direct-coupled throughout. A dc voltage

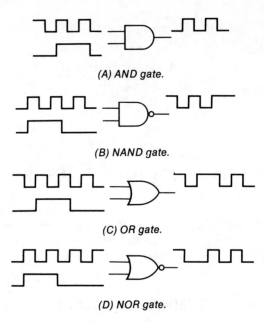

(A) AND gate.

(B) NAND gate.

(C) OR gate.

(D) NOR gate.

Fig. 3-1. Input and output gate waveforms.

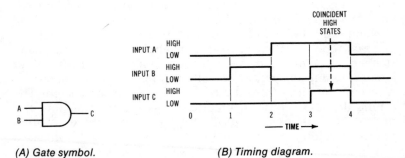

(A) Gate symbol.

(B) Timing diagram.

Fig. 3-2. Typical timing diagram for an AND gate.

applied at the input will maintain a corresponding output as long as the input voltage is applied. This is the basis of static troubleshooting procedures.

HALF ADDER

An AND gate is connected with its inputs in parallel with an XOR gate to form a basic adder, called a *half adder*, as shown in Fig. 3-4. A half adder has two input terminals, designated A and

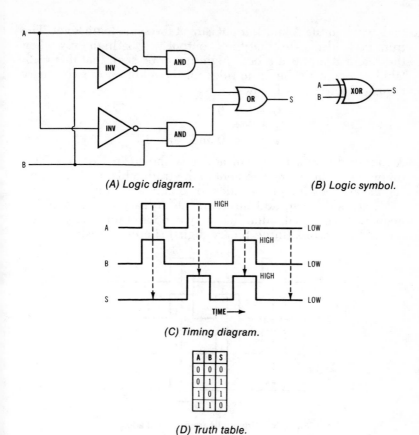

(A) Logic diagram.   *(B) Logic symbol.*

(C) Timing diagram.

A	B	S
0	0	0
0	1	1
1	0	1
1	1	0

(D) Truth table.

Fig. 3-3.   Diagrams of exclusive OR gate.

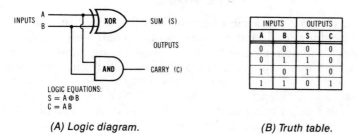

LOGIC EQUATIONS:
$S = A \oplus B$
$C = AB$

(A) Logic diagram.   *(B) Truth table.*

Fig. 3-4.   Configuration for a half adder.

B in Fig. 3-4. It has two outputs; they are designated S (the *sum* output) and C (the *carry* output). The truth table for a half adder (Fig. 3-4B) states that the S output will be high when the A and B inputs are at opposite logic levels, but that the S output will be low when the A and B inputs are at the same logic level. The truth table also states that the C output will be high only when the A and B inputs are both high. It can be seen that this truth table agrees with the basic rules for the addition of two binary digits, which are as follows:

$$0 + 0 = 0$$
$$1 + 0 = 1$$
$$1 + 1 = 0 \text{ and carry } 1$$

A timing diagram for a half adder is shown in Fig. 3-5. This timing diagram is a restatement of the truth table, which in turn represents a listing of the rules for binary addition.

A half adder can add any two bits (*binary digits*) simultaneously. Since a half adder has only two input terminals, it cannot add three bits (three binary digits) simultaneously.

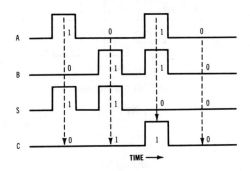

Fig. 3-5. Timing diagram for half adder.

FULL ADDER

Next, observe the logic diagram for a full adder in Fig. 3-6. The full-adder configuration consists of two half adders with an OR gate. A full adder forms the sum of two bits inputted at A and B. Also, if there happens to be a carry-in (C′) from a previous adder, the full adder will process the carry-in. Fig. 3-7A is the truth table for a full adder, and the same information is repeated in the form of a timing diagram in Fig. 3-7B. An example of a dual full adder is illustrated in Fig. 3-8. In this illustration, C_n corresponds to C′, C_{n+1} corresponds to C, and Σ (capital Greek letter sigma) corresponds to S.

As an example, suppose inputs A, B, and C′ in Fig. 3-6 are all driven high simultaneously. The sum output (S) will go high, and the carry output (C) will also go high. This is equivalent to the operation $1 + 1 + 1 = 11$ in binary arithmetic.

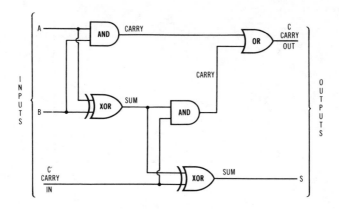

Fig. 3-6. Logic diagram of a full adder.

A	B	C'	S	C
0	0	0	0	0
0	0	1	1	0
0	1	0	1	0
0	1	1	0	1
1	0	0	1	0
1	0	1	0	1
1	1	0	0	1
1	1	1	1	1

(A) Truth table.

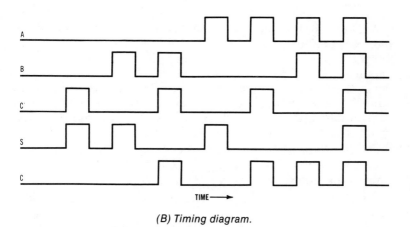

(B) Timing diagram.

Fig. 3-7. Function of a full adder.

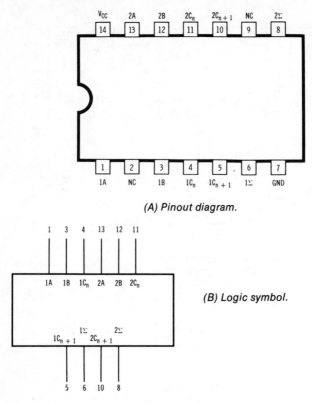

(A) Pinout diagram.

(B) Logic symbol.

Fig. 3-8. Type 93H183 dual full adder.

Adders may be arranged in cascade or in parallel. For example, two two-bit adders may be configured in parallel to add pairs of binary digits, as in the addition $11+10=101$. Still more parallel-connected adders are used to add larger binary numbers. These arrangements are called *parallel adders*. On the other hand, a *serial adder* performs the addition operation in sequential steps. Adders are used for multiplication; thus, 7×3 means that three sevens are to be added together or that seven threes are to be added together.

Logic Flow Diagram

A logic flow diagram for a full adder is shown in Fig. 3-9. This is a specialized diagram in which the A input is logic-high, the B input is logic-low, and the C' input is logic-high. The logic-high nodes (interconnections) in the configuration are indicated

62

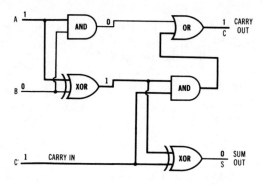

$$S = \overline{A}\overline{B}C' + \overline{A}B\overline{C}' + A\overline{B}\overline{C}' + ABC'$$
$$C = \overline{A}BC' + A\overline{B}C' + AB\overline{C}' + ABC'$$

Fig. 3-9. A specialized flow diagram.

by heavy lines; the logic-low nodes are indicated by thin lines. It is evident that each gate is responding to its inputs according to its truth table. The logic equation for the S output is written: $S=\overline{A}\overline{B}C'+\overline{A}B\overline{C}'+A\overline{B}\overline{C}'+ABC'$. This means that S goes logic-high if C' is high, OR if B is high, OR if A is high, OR if A, B, and C' are simultaneously high. The logic equation for C is written: $C=\overline{A}BC'+A\overline{B}C'+AB\overline{C}'+ABC'$. This means that C goes logic-high if B AND C' are high, OR if A AND C' are high, OR if A AND B are high, OR if A, B, and C' are high. (A configuration with two outputs must have two logic equations.)

Four-Bit Full Adder   •

A logic diagram for a four-bit full adder is shown in Fig. 3-10. This is a parallel adder in which a four-bit binary number, $A_1A_2A_3A_4$, is added to another four-bit binary number, $B_1B_2B_3B_4$. Note that although A_1 and B_1 are added simultaneously, A_2 and B_2 are added simultaneously, and so on, a small delay occurs before the appearance of the final sum because the carries must *ripple through* the network from the A_1B_1 section to the A_4B_4 section. The carry-out, C_0, is the last output to be finalized. This ripple-carry action results in *propagation delay,* and it places a limit on the speed with which the full adder can operate. As an illustration of ripple-carry action, note the following steps that are required in the formation of the final sum of the binary numbers 1101 and 1011.

$$\begin{array}{r} 1101 \\ +1011 \\ \hline \end{array}$$

(1)   10110 and 1 to carry

$$\begin{array}{r} 10110 \\ +1 \\ \hline \end{array}$$

(2)   10100 and 1 to carry

$$\begin{array}{r} 10100 \\ +1 \\ \hline \end{array}$$

(3)   10000 and 1 to carry

$$\begin{array}{r} 10000 \\ +1 \\ \hline \end{array}$$

(4)   11000 and 0 to carry (final sum)

Ripple Carry

This type of adder is called a *ripple-carry adder.* Ripple-carry action can be visualized to good advantage in the addition of binary 1111 and 0001 (decimal 15 and 1), as depicted in Fig. 3-11. Note that 1 cannot be added to 15 with a single propagation delay when a ripple-carry adder is employed. Instead, the first propagation delay involving the addition of A_1 and B_1 is followed by a second propagation delay involving the first carry. This is followed by a third propagation delay involving the second carry. In turn, this is followed by a fourth propagation delay involving the third carry.

Propagation delay is proportionally greater for 8-bit and 16-bit ripple-carry adders. Therefore, when speed of operation is an important consideration (as is generally the case), elaborated adder circuitry is utilized to reduce the propagation delay greatly. This topic is explained in a following chapter.

BASIC SUBTRACTER

The logic diagram for a half subtracter is shown in Fig. 3-12. A half subtracter is a unit or device that is capable of representing the difference between two numbers; it is usually restricted to operating with a subtrahend that has only one nonzero digit. A half subtracter is comparable to a half adder. The logic diagram for a full subtracter is shown in Fig. 3-13. The full subtracter is comparable to a full adder. It is an arrangement that provides for a borrow-in digit and a borrow-out digit.

Full subtracters are usually cascaded; the operation of a full

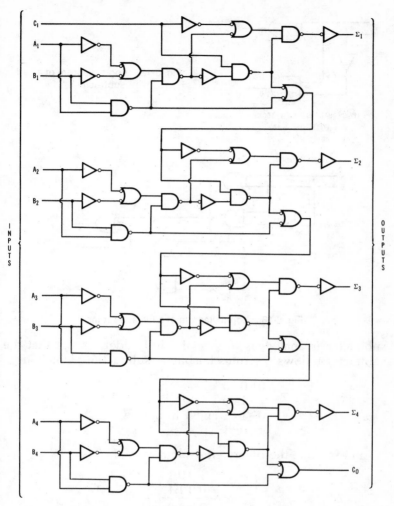

Fig. 3-10.   A four-bit parallel adder with ripple carry.

Fig. 3-11.   Addition of binary 1111 and binary 0001.

15
A_4 A_3 A_2 A_1
1   1   1   1

1
B_4 B_3 B_2 B_1
0   0   0   1

```
    1 1 1 1   AUGEND
  + 0 0 0 1   ADDEND
  ─────────
        1 1 1 0   FIRST SUM DIGIT
      1 1 1     SUCCESSIVE RIPPLE CARRIES
  ─────────
  1 0 0 0 0   COMPLETED SUM
```

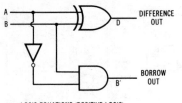

Fig. 3-12. Logic diagram of a half subtracter.

LOGIC EQUATIONS (POSITIVE LOGIC):

$$D = A \oplus B = A\bar{B} + \bar{A}B$$
$$B' = \bar{A}B$$

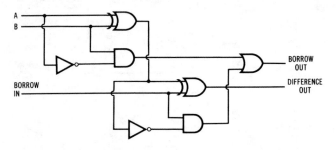

Fig. 3-13. Logic diagram of a full subtracter.

subtracter is analogous to that of a full adder, except that the subtracter follows the rules for binary subtraction, which are:

$$0-0=0$$
$$1-1=0$$
$$0-1=1 \text{ and 1 to borrow}$$
$$1-0=1$$

As an example, the binary equivalent of $9-3=6$ is:

$$\begin{array}{r} 1\ 0\ 0\ 1 \\ -0\ 0\ 1\ 1 \\ \hline 0\ 1\ 1\ 0 \end{array}$$

Full subtracters are used only to a minor extent in modern digital equipment, because an adder performs subtraction when supplemented by a controlled inverter, as explained next.

CONTROLLED INVERTER

A controlled inverter is an XOR gate, one input of which is connected to a data line and the other input of which is connected to a control line (Fig. 3-14). A binary digit (1 or 0) to be

66

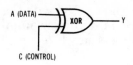

INPUTS		OUTPUT
A	C	Y
1	1	0
0	1	1
1	0	1
0	0	0

(A) Logic Symbol. *(B) Truth table.*

Fig. 3-14. Controlled inverter.

processed is applied to one input of the XOR gate, and a logic-high or a logic-low control voltage is applied to the other input of the gate. Controlled inverters are widely used in twos-complement adder/subtracters, as shown in Fig. 3-15. This arrangement will form the sum of $A_0A_1A_2A_3$ and $B_0B_1B_2B_3$, or it will form the difference between $A_0A_1A_2A_3$ and $B_0B_1B_2B_3$, depending on whether the control line is held logic-high or logic-low.

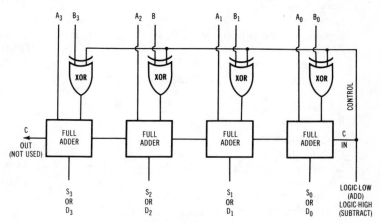

Fig. 3-15. Controlled-inverter adder/subtracter.

In order to understand Fig. 3-15, it is necessary to know how subtraction using twos complements is accomplished. The ones complement of a binary number is formed by changing all the ones to zeros and all the zeros to ones. For example, the ones complement of 1101 is 0010. The twos complement is formed by adding unity (one) to the ones complement. The twos complement of 1101 is 0010 + 1 = 0011. The use of twos complements for the subtraction of binary numbers can be shown most easily by means of the following examples. Example 1 shows the normal addition of two positive binary numbers, and Exam-

67

ples 2, 3, and 4 show subtraction (which is equivalent to adding negative numbers). Each example is shown first with decimal numbers and then with the equivalent binary numbers. (Equivalent decimal and binary numbers are listed in Table 3-1.)

EXAMPLE 1—TWO POSITIVE NUMBERS
Decimal Numbers:
$+ 4 + 2 = + 6$
Binary Numbers:

Enter 0100	0100
Enter 0010	0010
Find sum	0110

EXAMPLE 2—NEGATIVE NUMBER SMALLER THAN POSITIVE NUMBER
Decimal Numbers:
$+ 4 - 2 = + 2$
Binary Numbers:

Enter 0100	0100
Enter twos complement of 0010 = 1101 + 1 =	1110
Find sum	10010
Cast out the carry-out	0010

EXAMPLE 3—NEGATIVE NUMBER LARGER THAN POSITIVE NUMBER
Decimal Numbers:
$+ 2 - 4 = - 2$
Binary Numbers:

Enter 0010	0010
Enter twos complement of 0100 = 1011 + 1 =	1100
Find sum	1110
Because the answer is negative, find the twos complement of the sum; this is the absolute value of the answer	0001 + 1 = 0010

EXAMPLE 4—TWO NEGATIVE NUMBERS
Decimal Numbers:
$-4 -2 = -6$
Binary Numbers:

Enter twos complement of 0100 = 1011 + 1 =	1100
Enter twos complement of 0010 = 1101 + 1 =	1110
Find sum	11010
Cast out carry-out	1010
Find twos complement (because the answer is negative); this is the absolute value of the answer.	0101 + 1 = 0110

Table 3-1. Decimal and Binary Numbers

Decimal	Binary	
0		0000
1		0001
2		0010
3		0011
4		0100
5		0101
6		0110
7		0111
8		1000
9		1001
10		1010
11		1011
12		1100
13		1101
14		1110
15		1111
16	0001	0000
17	0001	0001
31	0001	1111
32	0010	0000

Now return to Fig. 3-15. When a controlled-inverter adder/subtracter is utilized, all positive numbers are entered into memories and registers without alteration. All negative numbers are entered into memories and registers in twos-complement form. When two binary numbers are to be added with the circuit of Fig. 3-15, the control line is held logic-low and the B bits then pass into the adders without inversion. (This can be verified by reference to the truth table in Fig. 3-14B.) Accordingly, the adders form the sum, $S_0S_1S_2S_3$, of the input binary numbers. On the other hand, when the two binary numbers are to be subtracted, the control line is held logic-high. The B bits then are inverted (complemented) before they pass into the adders. When the control line is driven logic-high, a 1 is applied to the carry-in input, and this 1 is added to the inverted binary number. Consequently, the twos complement of the number is formed. Accordingly, the adder/subtracter now functions to subtract the B bits from the A bits by the method shown in Example 2. Thus, $A_0A_1A_2A_3$ is the minuend, $B_0B_1B_2B_3$ is the subtrahend, and $D_0D_1D_2D_3$ is the difference between the binary numbers. Note that if the adder chain overflows, the carry-out is "cast out," or dropped, in accordance with the rules for subtraction by twos complements.

In Example 4, +2 was subtracted from −4. Suppose instead that −2 is to be subtracted from −4. In this case, −4 will be registered as 1100. Subtraction of −2 is the same as addition of +2, so the sum 1100 + 0010 = 1110 is found. The answer, being negative, is in twos-complement form. Its absolute value is 0001 + 1, or 0010.

TRUE/COMPLEMENT ZERO/ONE ELEMENT

The true/complement zero/one element shown in Fig. 3-16 passes the input data unchanged to the output when control input B is logic-low and control input C is logic-high. When both control inputs are driven logic-low, the input data is outputted in complementary form. When control input C is driven logic-low, a connection is provided to the carry-in terminal of the following full adder so that the adder will process the twos complement of the data that is inputted to the true/complement zero/one element.

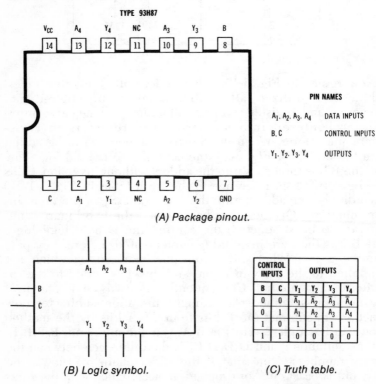

Fig. 3-16. A true/complement zero/one element.

Fig. 3-17 shows a logic diagram of a true/complement zero/one element. (Note that practically all digital-logic circuitry is direct-coupled [dc-coupled] throughout.)

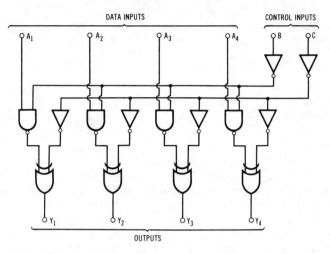

Fig. 3-17. Logic diagram for a true/complement zero/one element.

BASIC TROUBLESHOOTING PROCEDURES

Various faults and malfunctions occur in digital circuitry and produce trouble symptoms that require analysis. For example, if there is a solder bridge between adjacent pins on an IC, the two conductors are shorted together, and both have the same logic level even though the digital logic diagram may call for different levels. Conversely, a faulty bond (joint) in a signal path can produce an internal open circuit (inside the IC) that causes the associated output(s) to contradict the relations in the digital logic diagram. A solder bridge in the PC board circuitry can cause a "stuck at" trouble symptom—the conductor (node) might be stuck low and be unable to go high. Or, a short inside the IC might cause a stuck-high symptom—the node cannot be driven low.

Preliminary troubleshooting is usually accomplished to best advantage by use of a logic pulser and logic probe, as depicted in Fig. 3-18. A logic pulser is a miniature pulse generator that can be used to inject a single-shot pulse into a gate, for example. A logic probe is a pulse indicator; when the probe is applied at the output of a gate, an LED in the end of the probe indicates whether a pulse voltage is present or absent. For example, a

71

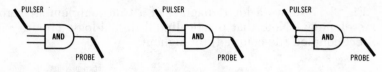

(A) Pulse in; normally no
pulse out.

(B) Pulse in; normally no
pulse out.

(C) Pulse in; normally
pulse out.

Fig. 3-18.   Use of digital logic probe and pulser.

three-input AND gate will normally have no output unless all three inputs are simultaneously pulsed high (Fig. 3-18). A multipin stimulus cable is used for convenient paralleling of input terminals for pulse tests.

Logic circuits often operate at high speed, and it may or may not be possible to troubleshoot a circuit with an unassisted logic probe. In such a case, troubleshooting is accomplished by supplementing the probe with a manual pulser so that the circuit can be operated by pulses manually injected one at a time. This is called *real time* troubleshooting, as contrasted to *clock time* troubleshooting. It is interesting to note that a single-shot pulse generator can be improvised by means of a $0.1\text{-}\mu\text{F}$ capacitor provided with test leads. One of the test leads is connected to ground, and the other is connected to V_{CC} (typically 5 volts) to charge the capacitor. Then, the lead to V_{CC} is disconnected and touched to the input terminal of concern. The capacitor discharges into the input terminal and injects a narrow test pulse. A logic probe, in turn, serves to indicate if a corresponding pulse is outputted from the circuit under test in accordance with the associated truth table.

A high-performance logic probe and logic pulser are illustrated in Fig. 3-19. The logic pulser can be set to inject single pulses or various trains of pulses. Effectively, this logic pulser is a miniature digital signal generator. The probe and pulser are connected to V_{CC} and ground as depicted in Fig. 3-20. A logic current tracer is useful for digital signal tracing in very low-impedance circuits. A large pulse of current in a low-impedance circuit is associated with a small pulse of voltage; if the signal voltage is too small to be detected by a logic probe, it can be easily indicated by a current-tracer test. A logic clip may be inserted over an IC for indication of the logic state at each pin. These testers are covered in greater detail subsequently.

As summarized in Chart 3-1, preliminary troubleshooting of a digital network is facilitated by "sizing up" the associated logic diagram. In other words, considerable time usually can be saved by evaluating the trouble symptoms with respect to the

72

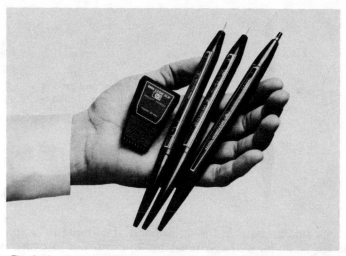

Fig. 3-19. (Left to right) logic clip, logic probe, logic pulser, current tracer. (*Courtesy Hewlett-Packard*)

Chart 3-1. "Sizing Up" Digital Logic Diagrams

Preliminary Considerations:
1. What is the basic function of the digital system?
2. List the functional subsections.
3. Is the malfunction solid (permanent), or is it intermittent?

Evaluation of Trouble Symptoms:
1. Note precisely what happens when the malfunction occurs.
2. How and when does the fault become observable?
3. Is the failure recurrent?
4. Do trouble symptoms appear in a significant sequence?

Progressive Evaluation:
1. Is there an instruction manual for the digital system?
2. Do you have a maintenance manual?
3. Is a service index with troubleshooting charts available?
4. Do you have a file of case histories?

Planning a Preliminary Approach:
1. Can you work up a system-down flow diagram?
2. Note all of the functional subsections that could possibly be involved.
3. Consider signal-injection and signal-tracing tests that can be made to narrow down the possible trouble area.
4. Have all obvious possibilities been checked, such as loose connections, low line voltage, and "cockpit error" (incorrect assumptions or conclusions by the technician)?
5. Is a similar digital system available in normal working condition for making comparative tests?

Fig. 3-20. Power and ground connections for logic probe or logic pulser.

network function(s). On the other hand, if a "shotgun" approach is used, each IC package must be "buzzed out" in turn with respect to its pertinent truth table. If a unit contains dozens or scores of IC packages, the probability of locating the fault within a reasonable length of time is very poor. Efficient troubleshooting requires understanding of circuit operation so that various devices and sections can be eliminated from suspicion at the outset.

Chapter
4

Additional Gate Types and Logic Families

Several basic types of logic gates, such as AND, OR, NAND, NOR, XOR, and XNOR, have been introduced in earlier chapters. In this chapter, several additional types of gate functions will be described. Also, the characteristics of some of the important families of logic devices will be discussed.

EXPANDABLE GATES

When examining logic diagrams, the technician will occasionally encounter *expandable gates*, such as the one shown in Fig. 4-1. The expander input is marked X in this example. An expandable gate is defined as a logic gate with provision for increasing the number of inputs by the addition of a logic block. The X input in Fig. 4-1 is a connection to the output section of the OR gate in the device. If the X input is pulsed, an output pulse occurs, just as if AB, CD, EF, or GHI were pulsed. This function is stated in the logic equation. Up to four external gates may be channeled into the X input.

MAJORITY LOGIC

Majority logic is also called *threshold logic*. It is a form of logic that makes decisions based on the number of inputs at a given logic level. A majority-logic gate with five inputs is shown in Fig. 4-2. The gate output goes logic-high whenever any three or more of the inputs are driven logic-high. As shown in Fig. 4-3, an XNOR gate is included in the conventional majority-logic

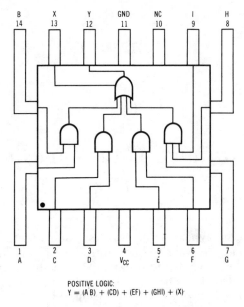

POSITIVE LOGIC:
$Y = (A\,B) + (CD) + (EF) + (GHI) + (X)$

Fig. 4-1. An expandable 2-2-2-3–input AND-OR gate.

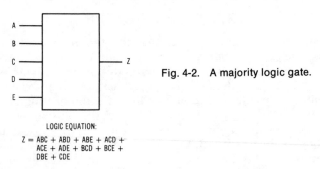

Fig. 4-2. A majority logic gate.

LOGIC EQUATION:
$Z = ABC + ABD + ABE + ACD +$
$ACE + ADE + BCD + BCE +$
$DBE + CDE$

package. One input of the XNOR gate is connected to an IC pin
(W in Fig. 4-3). If input W is tied to a logic-high level, the arrangement functions as a five-input majority gate. Its output
goes logic-high whenever any three or more of its inputs are
driven logic-high. If input W is tied to a logic-low level, the
arrangement functions as a "minority-logic gate." Its output will
be logic-low whenever three or more of its inputs are logic-
high. The arrangement will operate as a three-input AND gate
when W is tied to a logic-high level and two of the other input
pins are tied to a logic-low level (Fig. 4-4). When W is tied to a

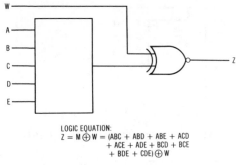

LOGIC EQUATION:
$$Z = M \oplus W = (ABC + ABD + ABE + ACD + ACE + ADE + BCD + BCE + BDE + CDE) \oplus W$$

Fig. 4-3. Majority logic gate with XNOR gate.

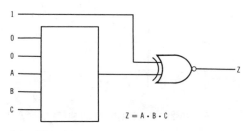

$Z = A \cdot B \cdot C$

Fig. 4-4. Majority-logic implementation of AND function.

logic-high level and two of the other pins are also tied to a logic-high level (Fig. 4-5), the arrangement operates as a three-input OR gate. Implementations of NAND and NOR gates are similarly obtained, as shown in Figs. 4-6 and 4-7.

To implement a two-input OR gate, W is tied to a logic-high level, two of the other inputs are also tied to a logic-high level, and another of the inputs is tied to a logic-low level. To implement a three-input majority-logic gate, the W input is tied to a logic-high level, one of the other inputs is also tied to a logic-high level, and another of the inputs is tied to a logic-low level. When any two of the remaining inputs are driven logic-high, the output will go logic-high.

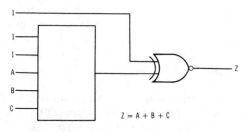

$Z = A + B + C$

Fig. 4-5. Majority-logic implementation of OR function.

77

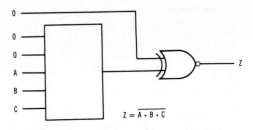

Fig. 4-6. Majority-logic implementation of NAND function.

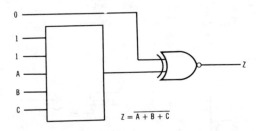

Fig. 4-7. Majority-logic implementation of NOR function.

WIRE-AND/OR CONFIGURATIONS

Wire-AND/OR configurations are often encountered in logic diagrams. The output transistor of an inverter or a NAND gate may be designed with an open collector (Fig. 4-8). In other words, the collector of the output transistor in the device is connected to the output terminal and to no other internal circuit elements. Open-collector design is indicated by an asterisk (*) at the output terminal as in Fig. 4-8A. Because the collector is open, the device is not functional until an external *pull-up resistor* is connected to the output terminal, as shown in Fig. 4-9A. A pair of open-collector inverters may be utilized in a wire-AND configuration as shown in Fig. 4-9B. This arrangement uses a common pull-up resistor for both of the open-collector inverters. If both inputs (A and B) are low, both transistors are cut off, and both C and D are high. If either input (A or B) is high, one transistor is saturated (C or D goes low), and the common point of C, D, and Y is pulled low. Thus, if both inputs are low, the output (Y) is high; if either or both inputs are high, the output is low. These relationships are summarized in the truth table in Fig. 4-9B. Notice that the columns for C, D, and Y form the truth table for the AND function. Thus, the wired connection of C and D serves to AND the transistor outputs; this connection provides a *wired* AND, or wire-AND, function.

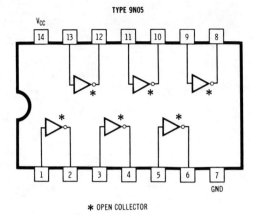

OPEN COLLECTOR

(A) Hex inverter package.

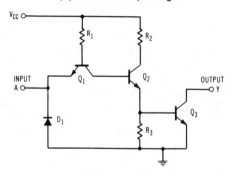

(B) Circuit of one inverter.

Fig. 4-8. Example of inverter with open-collector output.

The overall logic equation for the circuit in Fig. 4-9B is $Y = \overline{A} \cdot \overline{B}$. Observe that if the A and B inputs are negated by means of series inverters, the logic equation becomes $Y = A \cdot B$. Then, ANDing action is obtained in the same format as for a conventional AND gate.

Observe also that if the A and B inputs in Fig. 4-9B are not negated and if the output from the wire-AND circuit is complemented by means of a series inverter, the logic equation becomes $Y = A + B$, in accordance with DeMorgan's laws. That is, we write $Y = \overline{\overline{A} \cdot \overline{B}} = A + B$, and ORing action is provided in the same format as for a conventional OR gate. Wire-AND/OR configurations find extensive application in logic systems that require ANDing or ORing of a comparatively large number of in-

79

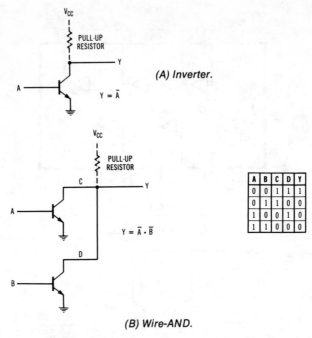

(A) Inverter.

$Y = \bar{A}$

$Y = \bar{A} \cdot \bar{B}$

(B) Wire-AND.

A	B	C	D	Y
0	0	1	1	1
0	1	1	0	0
1	0	0	1	0
1	1	0	0	0

Fig. 4-9.   Use of external pull-up resistors.

puts. Fig. 4-10 shows some logic gates implemented with open-collector inverters.

I²L GATES

The abbreviation I²L stands for *Integrated Injection Logic.* This logic family combines two desirable properties: simple fabrication and high-speed operation. As shown in Fig. 4-11A, the basic I²L gate structure is an inverter that is implemented as a multicollector transistor. A constant-current source derived from a pnp transistor supplies base drive to the npn multi-collector injection element. The basic function of this gate is inversion; however, a negated-AND (NOR) function is provided by wire-AND (wire-OR) circuitry.

A logic symbol for an I²L gate is shown in Fig. 4-11B. The multicollector outputs are indicated as separate output lines, with a single input line to the inverter gate. Multiple inputs to the gate are wire-ANDed together, as shown. The illustrated gate has a fan-out of three. Additional outputs can be provided for a fan-out of four or five. (Fan-out denotes the number of loads that can be driven.)

80

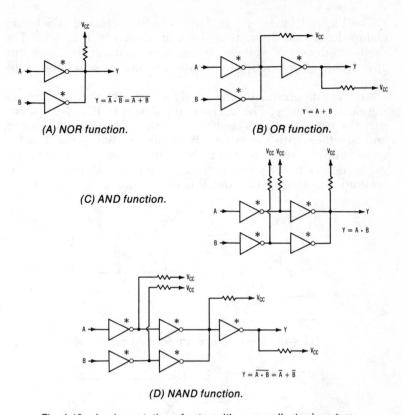

(A) NOR function.

(B) OR function.

(C) AND function.

$Y = A \cdot B$

(D) NAND function.

Fig. 4-10. Implementation of gates with open-collector inverters.

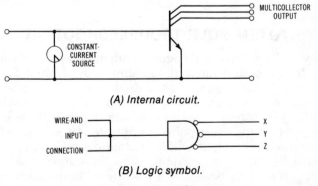

(A) Internal circuit.

(B) Logic symbol.

Fig. 4-11. Basic I²L gate.

81

The basic I²L gate may be improved by inclusion of Schottky clamp diodes on the output lines, as shown in Fig. 4-12. The diodes reduce the voltage swings and increase the operating speed; they also function as clamps, preventing saturation of the transistor.

An I²L gate eliminates the need for resistors, thereby increasing device density. The latch circuit shown in Fig. 4-13 is simple; the gates are cross-connected in the conventional manner, and together with the A and B inputs are wire-ANDed at the input. Simplification of inner connections contributes to propagation delays in the range between 1 and 5 ns. If the diodes are omitted, the propagation delay is on the order of 10 ns.

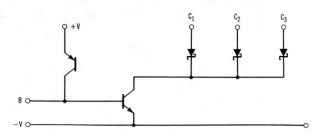

Fig. 4-12.   Basic I²L gate with Schottky clamp diodes.

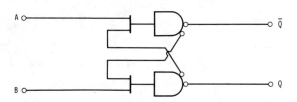

Fig. 4-13.   An I²L latch circuit.

TOTEM-POLE TROUBLESHOOTING

Many TTL gates are designed with totem-pole output circuitry. In this arrangement, the emitter of one output transistor is connected to (or coupled via a diode to) the collector of the other output transistor. The troubleshooter must consider the error effect of a short between two outputs when these outputs attempt to pull the same point to opposite states (Fig. 4-14). In this situation, the output which is attempting to pull the node high will be supplying current, while the output which is attempting to pull the node low is a saturated transistor to ground (Fig. 4-14B) and will be sinking the current. Therefore, the

82

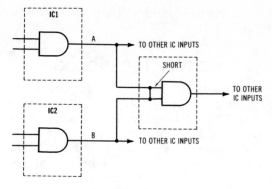

(A) Logic diagram.

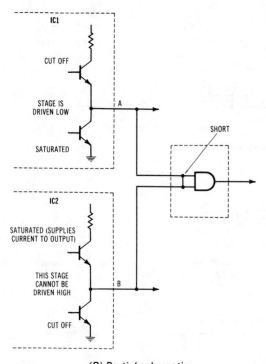

(B) Partial schematic.

Fig. 4-14. Short circuit between outputs of two AND gates with totem-pole output circuits.

83

saturated transistor to ground will pull the node to a low state ("stuck at" low).

MOSFET TECHNOLOGY

The acronym MOSFET stands for *Metal-Oxide Semiconductor Field-Effect Transistor*; MOS technology is an important supplement to TTL technology. Unipolar (field-effect) transistors are grouped into *junction-gate* types (JFETs) and *insulated-gate* types (IGFETs or MOSFETs). They are subgrouped into n-channel and p-channel types. Some representative FET symbols are shown in Fig. 4-15.

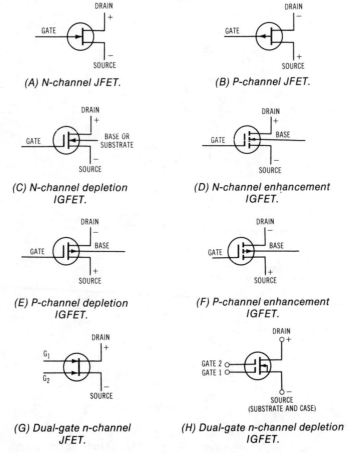

(A) N-channel JFET.

(B) P-channel JFET.

(C) N-channel depletion IGFET.

(D) N-channel enhancement IGFET.

(E) P-channel depletion IGFET.

(F) P-channel enhancement IGFET.

(G) Dual-gate n-channel JFET.

(H) Dual-gate n-channel depletion IGFET.

Fig. 4-15.   Examples of symbols for field-effect transistors.

As a practical troubleshooting note, insulated-gate transistors (IGFETs) are susceptible to damage by static electricity. Accordingly, suitable handling procedures must be observed when an IGFET is replaced in a circuit. Observe also that some MOSFETs operate with a negative power supply and others operate with a positive power supply.

MOS Integrated Circuits

Digital MOS ICs employ enhancement-type transistors almost exclusively because of the comparatively low power consumption of this type of transistor. An enhancement device is normally cut off until a sufficiently large forward-bias gate voltage is applied. Note also that the *complementary MOS* (CMOS) configuration is used extensively in digital circuitry. As shown in Fig. 4-16, a p-channel MOS transistor is connected in series with an n-channel MOS transistor. This arrangement is analogous to the TTL totem-pole configuration. When the input to the CMOS stage is logic-low, the p-channel transistor conducts, and the n-channel transistor is cut off. Conversely, when the input to the CMOS stage is logic-high, the p-channel transistor cuts off, and the n-channel transistor conducts.

The MOS inverter, depicted in Fig. 4-17, is a basic MOSFET logic element. This arrangement uses p-channel devices. Note that Q_1 approximates a constant-current load. When the data

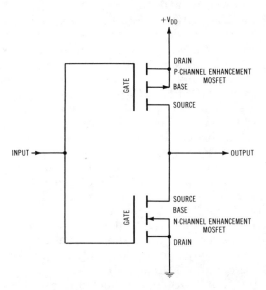

Fig. 4-16. Complementary MOS stage.

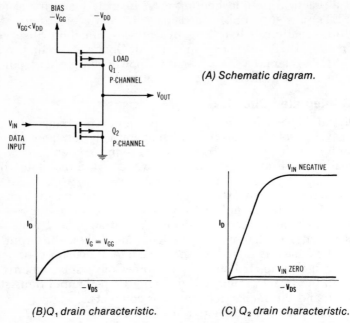

(A) Schematic diagram.

(B) Q_1 drain characteristic.

(C) Q_2 drain characteristic.

Fig. 4-17. Basic MOS inverter.

input is driven negative, the output is near ground potential (inverter action). The MOS inverter is an example of *static* device design. A circuit made up of static devices does not require a clock signal. On the other hand, dynamic devices, such as dynamic MOS logic gates, require clocking. Static MOS logic is utilized in memories, and dynamic MOS logic is also employed.

Complementary MOSFET (CMOS) gates are based on the CMOS inverter. The configuration for a three-input CMOS NAND gate is shown in Fig. 4-18. Observe that the pull-up MOSFETs are p-channel type, whereas the pull-down MOS-FETs are n-channel type. Thus, this is a complementary configuration. When inputs A, B, and C are driven logic-high simultaneously, the pull-up MOSFETs are cut off, and the pull-down MOSFETs are conducting. In turn, the output terminal has a low impedance to ground, and the output state is logic-low (NAND response). Note that Q_4, Q_5, and Q_6 are connected in series, whereas Q_1, Q_2, and Q_3 are connected in parallel. Accordingly, if only one of the inputs is logic-low, the output will be logic-high (NAND action).

A three-input CMOS NOR gate is shown in Fig. 4-19. Observe that the pull-up MOSFETs are connected in series, whereas the

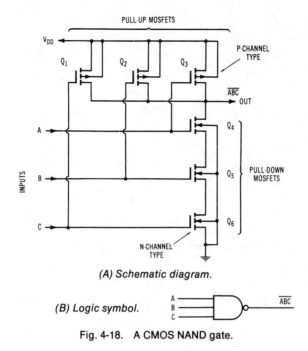

(A) Schematic diagram.

(B) Logic symbol.

A
B
C
$\overline{ABC}$

Fig. 4-18. A CMOS NAND gate.

pull-down MOSFETs are connected in parallel. If one input is driven logic-high, the pull-up circuit is cut off, and the pull-down circuit conducts. In turn, the output has a low impedance to ground, and its state is logic-low (NOR response). When all three inputs are driven logic-low, the pull-up circuit conducts, but the pull-down circuit is cut off. In turn, V_{DD} is applied to the output terminal, and its state is logic-high (NOR action).

Dynamic MOS

Dynamic MOS circuitry samples the incoming data periodically in response to clock pulses. The clock signal is typically a 1-MHz square-wave or pulse voltage. Dynamic MOS is characterized by low power consumption compared to static MOS. In the circuit of Fig. 4-20, voltage levels are stored on capacitances; clock pulses ϕ_1 and ϕ_2 (two-phase clock signals) serve to "refresh" charges that tend to decay. In other words, the gates of Q_1 and Q_3 are pulsed with a comparatively short "on" time; in turn, power consumption is considerably reduced compared to that of a static gate.

Observe that two-phase clock signals are employed in Fig. 4-20. The input V_{in} goes to zero volts at t_0, causing Q_2 to cut off.

87

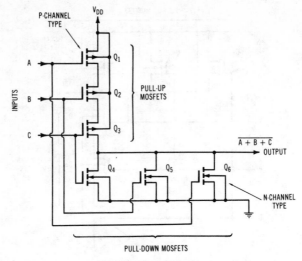

(A) Schematic diagram.

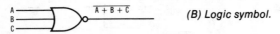

(B) Logic symbol.

Fig. 4-19. A CMOS NOR gate.

At the same time, ϕ_1 goes negative; this turns on Q_1, and C_1 charges to V_{DD}. Next, ϕ_2 turns on Q_3 at t_1. The conduction of Q_3 causes the charge that is on C_1 to be transferred to C_2. Almost the full voltage is transferred because C_1 is much larger than C_2. Note that Q_2 is turned on at t_2 by V_{in}; the conduction of Q_2 causes C_1 to partially discharge. However, this partial discharge does not affect C_2 at this time because Q_3 is cut off. Then, when ϕ_1 cuts off Q_1 at t_3, a partial charge remains on C_1. Since Q_2 is still held in conduction by V_{in}, C_1 discharges. Finally, at t_4, ϕ_2 turns on Q_3; then C_2 discharges to ground via Q_2.

In summary, during one clock cycle (from t_0 through t_4), drain-current demand occurs only from t_2 to t_3 (due to the fact that Q_1 and Q_2 are conducting simultaneously) and when C_1 charges at t_0. Note that drain-current demand occurs only when the ϕ_1 pulse is present.

The circuit in Fig. 4-20 is termed a *ratioless* configuration because the output level does not depend on the resistance ratio of Q_1 and Q_2.

Ratioless gates are elaborated inverter arrangements. Fig. 4-21 shows NAND and NOR functions based on the ratioless dynamic inverter. In Fig. 4-21A, NOR action is obtained by

88

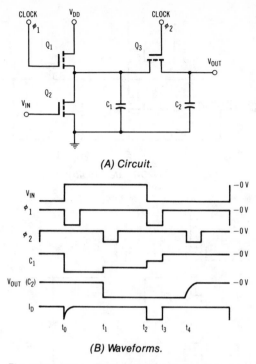

(A) Circuit.

(B) Waveforms.

Fig. 4-20. Basic dynamic MOS ratioless inverter.

means of Q_3 connected in parallel with Q_2. In other words, the basic inverter is provided with two inputs connected in parallel. To obtain NAND action, another input gate is connected in series with Q_2 (Fig. 4-21B). Accordingly, both inputs must be driven logic-high simultaneously to obtain a logic-low output. Circuit action is essentially the same as for the ratioless inverter arrangement, and the charges on the capacitors tend to decay. Therefore, the gates must be clocked in order to "refresh" the stored charges.

Ratioless-Powerless Inverter

The *ratioless-powerless inverter* in Fig. 4-22 consumes only reactive power, which is obtained from the two-phase clock pulses. This is a form of dynamic MOS circuitry that differs from the ratioless-inverter type of circuitry in that no dc power supply is employed. With reference to Fig. 4-22, at time t_0 C_1 is charged, C_2 and C_3 are discharged, and the input terminal is negative (logic-high). The input voltage goes to zero at t_1; at the same time, clock ϕ_1 turns on Q_1 and Q_2. In turn, C_1 discharges

89

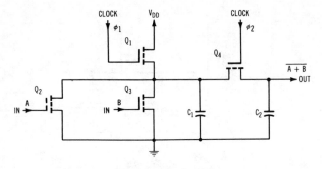

(A) NOR gate circuitry.

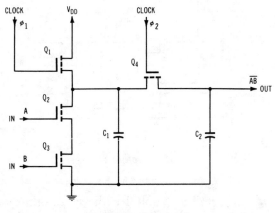

(B) NAND gate circuitry.

Fig. 4-21. Basic dynamic ratioless two-phase gates.

via Q_1, with the result that Q_3 is cut off. Accordingly, C_2 charges with Q_2 on and Q_3 off. The stored charge on C_2 is transferred to C_3 at t_2 in response to turn-on of Q_4 by ϕ_2.

Next, the input drive goes logic-high at t_3, and ϕ_1 turns on Q_1 and Q_2. Consequently, C_1 charges, with the result that Q_3 is turned on. Although both Q_2 and Q_3 are on at this time, C_2 cannot discharge because the source and the drain of Q_2 are at the same potential. Observe that Q_1 turns off at t_4; this occurs at the end of the ϕ_1 pulse. At this time, C_1 remains charged, Q_3 is on, and Q_2 is off. Accordingly, C_2 discharges to ground via Q_3.

Other more elaborate dynamic-logic circuits require up to four clock phases. Ratioless-powerless circuitry has the advantage of minimum power consumption, but at the expense of increased circuit complexity.

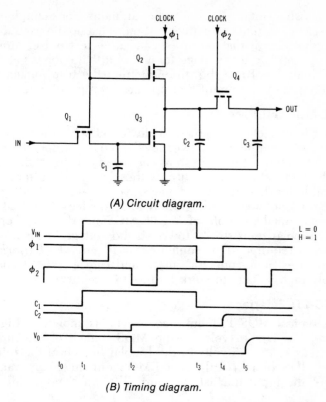

(A) Circuit diagram.

(B) Timing diagram.

Fig. 4-22. Basic dynamic two-phase ratioless-powerless inverter.

Negative Logic Level

It was noted in the preceding discussion that the gate input is driven negative for a logic-high input. This is opposite to TTL gate operation in which the gate input is driven positive for a logic-high input. In a circuit that uses negative supply voltages, confusion may arise concerning the functioning of the circuit. Most digital IC manufacturers have adopted the following convention:

1. A logic-high level is defined as the greatest potential, whether positive or negative, with respect to ground potential.
2. A logic-low level is defined as the potential nearest ground, whether positive or negative.

This is a matter of importance to the digital troubleshooter, not only from the viewpoint of reading logic diagrams, but also in choosing suitable digital test equipment. For example, a logic pulser that outputs a positive test pulse cannot be used to check a digital circuit that requires a negative test pulse. Moreover, the appropriate test voltage for a circuit that operates at a negative logic level may be different from the test voltage that is appropriate for a circuit that operates at a positive logic level.

TTL-MOS Interface

An interface is a point or a device at which a transition is made between media, power levels, modes of operation, or related functions. Fig. 4-23 shows an interface circuit that accepts TTL digital pulses and outputs these pulses in a form that is suitable for driving MOS devices. Note that a TTL-MOS interface does not necessarily employ MOS devices. Fig. 4-24 shows a commercial example of a TTL-MOS interface that employs only bipolar transistors. However, the voltage levels are translated as required. This configuration is termed a *buffer/driver*. It features open-collector output and can operate at comparatively high voltage to work into MOS circuitry.

MOS-TTL Interface

A typical MOS-TTL interface circuit is shown in Fig. 4-25. Here, V_{DD} is -14 volts in the MOS branch of the interface, whereas V_{CC} is $+5$ volts in the bipolar branch of the interface. The MOS branch utilizes a PMOS enhancement transistor. Thus, the input transistor is normally "off," and a sufficiently

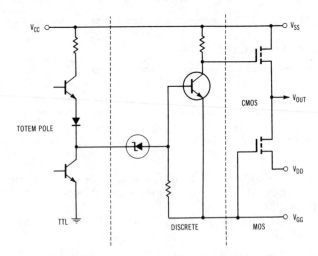

Fig. 4-23.   Example of a TTL-to-MOS interface circuit.

TYPE 9N06

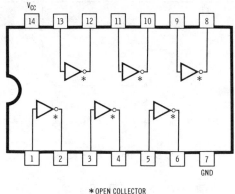

* OPEN COLLECTOR
POSITIVE LOGIC: $Y = \overline{A}$

(A) Package pinout of hex device.

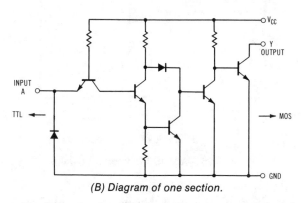

(B) Diagram of one section.

Fig. 4-24. An open-collector hex inverter buffer/driver
TTL-MOS interface.

large gate input voltage must be applied to turn it "on." The gate-voltage swing is typically from −5 to −14 volts. In accordance with the definition stated previously, the logic-high level is −14 volts, and the logic-low level is −5 volts. It follows that the technician must be alert to the difference between TTL voltage values and polarities and MOS voltage values and polarities when troubleshooting digital systems that operate logic-family interfaces.

MOS Flip-Flop

A basic MOS flip-flop circuit is shown in Fig. 4-26. (High-capacity memories are built up from MOS flip-flops.) A flip-flop

93

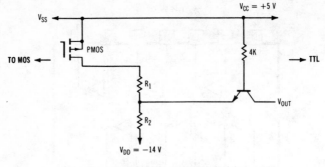

(A) Diagram of basic circuit.

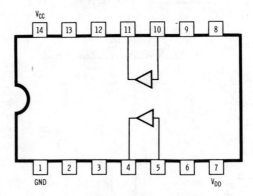

(B) Example of IC package.

Fig. 4-25.　A MOS-TTL interface circuit.

is also called a bistable multivibrator, or trigger circuit. It is a two-stage multivibrator circuit that has two stable states. In one state, the first stage is conducting and the second stage is cut off. In the other state, the second stage is conducting and the first stage is cut off. A trigger signal changes the circuit from one state to the other. Various types of flip-flops are described in Chapter 5.

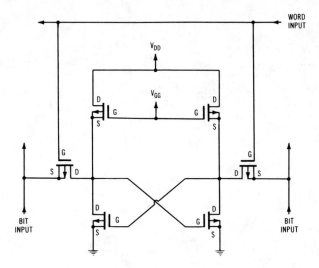

Fig. 4-26. Basic MOS flip-flop configuration.

Chapter
5

Flip-Flops and Clock Circuitry

Various forms of relaxation oscillators are encountered in digital logic diagrams. A conventional multivibrator (Fig. 5-1), for example, can be used as a *clock oscillator* in a scanner-monitor radio. This is a free-running multivibrator that generates a square-wave output with a repetition rate of several hertz. It is an example of an *astable* multivibrator. We will also encounter *monostable* multivibrators. A monostable multivibrator is sometimes called a *one-shot* or *single-shot* multivibrator, because when it is triggered it outputs a single pulse.

The *bistable* multivibrator, commonly termed a *flip-flop*, is widely used in digital circuitry. A flip-flop (FF) is a multivibrator that has two stable states; its output will remain logic-high until it is appropriately triggered, whereupon its output will go logic-low. Then, its output will remain logic-low until it is again triggered, whereupon its output goes logic-high. There

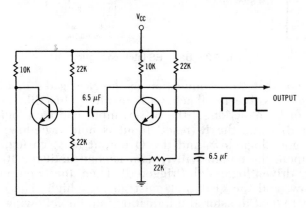

Fig. 5-1. Typical free-running multivibrator.

97

are various groups and subgroups of flip-flops, and what is normal operation for one class of flip-flops may represent a malfunction for another class. Therefore, the digital troubleshooter needs to have a good understanding of flip-flop operation.

BASIC SET-RESET LATCH

A latch is a bistable multivibrator used for temporary storage of binary information in a digital system. Thus, a latch is a temporary *memory* that can be erased, or *reset,* as required. The latch stores data for a specified time, and then unloads it. Two NOR gates are used in a basic set-reset (RS) latch configuration, as shown in Fig. 5-2. Although the terms *latch* and *flip-flop* are often used interchangeably, there is a technical distinction between them. A latch is an asynchronous device in the strict sense of the term; a latch can be triggered at any arbitrary time. On the other hand, a flip-flop is a synchronous device, strictly speaking; a flip-flop can be triggered only upon arrival of a clock pulse.

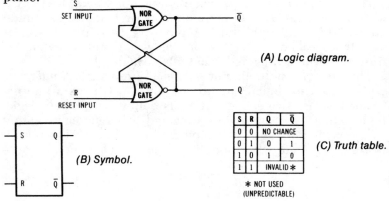

(A) Logic diagram.

(B) Symbol.

(C) Truth table.

S	R	Q	Q̄
0	0	NO CHANGE	
0	1	0	1
1	0	1	0
1	1	INVALID *	

* NOT USED
(UNPREDICTABLE)

Fig. 5-2. Basic RS latch configuration.

Observe that cross-connection of the NOR gates in Fig. 5-2 provides positive feedback with resulting bistable multivibrator, or latch, action. If the S (set) input is driven logic-high (triggered) while the R (reset) input is held logic-low, the Q̄ output goes logic-low, and the Q output must go logic-high. Thereupon, the circuit will remain latched until the R (reset) input is driven logic-high (triggered). Then, the Q output goes logic-low, and the Q̄ output must go logic-high. The outputs thereby *unload* the stored information into the following circuit (not shown).

RACES

Observe in Fig. 5-2 that the R and S inputs of an RS latch must not be driven logic-high simultaneously. This is a forbidden operating condition, because the output states would be unpredictable. If R goes high slightly before S goes high, Q goes low, but if S goes high before R goes high, Q goes high. Technicians refer to this condition as a *race*. Since various digital malfunctions and trouble symptoms are caused by the development of race situations, it is desirable to take a closer look at the causes and effects of races.

With reference to Fig. 5-3, a race at the inputs to an AND gate can result in outputting a spurious pulse, or *glitch*. If a glitch has appreciable width, the digital circuitry cannot distinguish between it and a true digital pulse, and a malfunction results. A race condition occurs for the AND gate when input A goes from logic-high to logic-low as input B goes from logic-low to logic-high. If the fall time of the A waveform is comparatively slow, the A and B inputs will momentarily be logic-high simultaneously. In turn, a narrow pulse (glitch) occurs at the output.

In the example of Fig. 5-3, the glitch could be avoided by improving the fall time of the A waveform, or by including a buffer in series with the B input. A buffer introduces a small propagation delay so that the B input does not go logic-high until the A input has gone logic-low. Logic diagrams sometimes show buffers that function merely to insert a needed propagation delay in order to resolve a race problem.

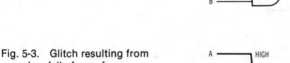

Fig. 5-3. Glitch resulting from slow fall of waveform.

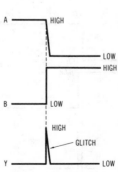

SYSTEM PROPAGATION DELAY

Races and glitch generation can occur even when waveforms are ideal. The A and B input waveforms in Fig. 5-4 are ideal; the rise and fall times are zero. However, the trailing edge (fall) of the B waveform is slightly late due to propagation delay in a previous part of the digital system. As a result, the A and B inputs are logic-high simultaneously until the B input attains the logic-low state. The production of a glitch could be avoided by including a buffer in the A input line. Sometimes two buffers are placed in series to provide the necessary amount of propagation delay.

NOISE GLITCHES

Race glitches should not be confused with noise glitches when troubleshooting a malfunctioning digital system. A noise glitch (Fig. 5-5) may be tracked down to a defective power supply, for example. An oscilloscope with extended high-frequency response and triggered sweep is the most useful instrument for analyzing and tracing either race glitches or noise glitches in digital circuitry.

RS LATCH APPLICATION

A typical application for an RS latch is shown in Fig. 5-6. In this case, the latch is used in a digital bowling game. A Darlington phototransistor responds to light-intensity changes and

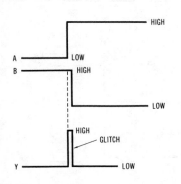

Fig. 5-4.   Glitch due to system propagation delay.

100

develops a low resistance to ground when brightly illuminated. In turn, the Q output, which is normally logic-low, goes logic-high when a ball passes over the phototransistor. The output from the latch is displayed by a scoreboard indicator. Then, with both inputs logic-low, no change occurs in the output. However, when the reset switch is pressed, Q goes logic-low.

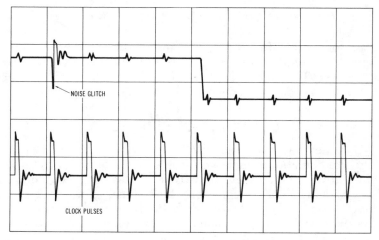

Fig. 5-5. Example of a noise glitch.

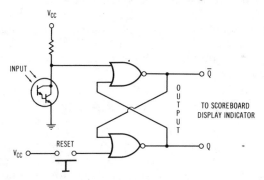

Fig. 5-6. An application for an RS latch.

D LATCH

Observe the D-latch arrangement shown in Fig. 5-7. The S input is driven by the incoming data, and the R input is driven by inverted (complemented) data. The result is that when the data is logic-high, Q goes logic-high and $\overline{Q}$ goes logic-low. Conversely, when the data is logic-low, Q goes logic-low and $\overline{Q}$

101

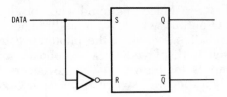

Fig. 5-7. Basic D-latch configuration.

goes logic-high. It is obviously impossible to drive S and R logic-high simultaneously, or to drive S and R logic-low simultaneously. The chief advantage of the D latch is that the race condition cannot occur in any circumstance.

CLOCKED RS FLIP-FLOP

Consider next the clocked RS flip-flop depicted in Fig. 5-8. This configuration is more versatile than the basic RS latch. The clocked RS flip-flop includes two AND gates connected to the R and S inputs, with a common clock signal. The clock signal functions to *enable* or to *disable* both AND gates simultaneously. The R and S inputs can change the state of the latch only while the clock signal is logic-high. When the clock signal is logic-low, the flip-flop is unresponsive to the states of the R and S inputs. The clock signal is said to provide a *window*. This window is open only while the clock signal is logic-high. One advantage of clocking is that the R and S lines may be used to trigger circuits other than the RS flip-flop for half of the total elapsed time.

PRESET AND CLEAR INPUTS

A clocked flip-flop operates in synchronism with other circuits in a total digital system. Synchronizing permits precise data processing sequences at high speed. Note that *preset* and *clear* inputs are also provided in the clocked flip-flop configuration (Fig. 5-8). These inputs are used to set or reset the flip-flop independently of the data and clock inputs. In other words, the preset input can be driven logic-high at any time, and the Q input will go logic-high at that time. Conversely, the clear input can be driven logic-high at any time, and the Q input will go logic-low at that time. Because the preset and clear inputs override the data and clock inputs, it is said that the preset and clear functions are asynchronous. The set and reset functions, on the other hand, are synchronous.

102

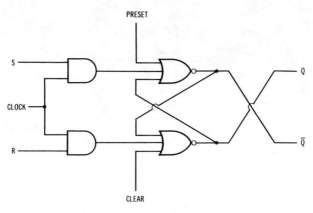

Fig. 5-8. Clocked RS flip-flop.

When a digital system is started up (power is applied), the starting states of the RS flip-flops cannot be predicted. Thus, one flip-flop might have its Q output logic-high, whereas another flip-flop might have its Q output logic-low. However, if the preset inputs are automatically pulsed logic-high at the outset, the flip-flops will be *initialized* with all of their Q outputs logic-high.

A race condition can arise in a clocked RS flip-flop circuit, just as in an RS latch circuit. Consequently, the basic clocked RS flip-flop is often elaborated to eliminate the possibility of a race under any operating conditions.

CLOCKED D FLIP-FLOP

The basic D latch configuration was shown in Fig. 5-7. The same principle is utilized with the clocked RS flip-flop (Fig. 5-8) to form the clocked D flip-flop arrangement shown in Fig. 5-9. With the (R) input always inverted with respect to the (S) input, a race condition cannot occur with respect to these inputs. However, if the preset input and the clear input are driven logic-high simultaneously, a race situation will arise. Observe the timing diagram for the clocked D flip-flop in Fig. 5-10A. While the clock is logic-high, a 1 level on the D (data) line can pass through the (S) gate, and the Q output goes logic-high. The Q output then remains logic-high unless the data line goes logic-low and the clock line goes high at the same time.

A clocked D flip-flop cannot be reset until such time as the data line is logic-low while the clock goes high. This is the

103

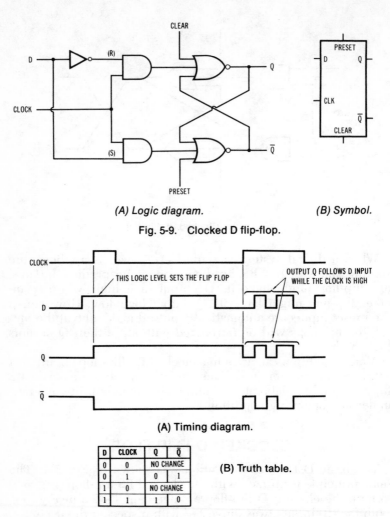

(A) Logic diagram.   *(B) Symbol.*

Fig. 5-9. Clocked D flip-flop.

(A) Timing diagram.

D	CLOCK	Q	Q̄
0	0	NO CHANGE	
0	1	0	1
1	0	NO CHANGE	
1	1	1	0

(B) Truth table.

Fig. 5-10. Operation of clocked D flip-flop. (*Courtesy Hewlett-Packard*)

previously mentioned window action that provides flexibility in application inasmuch as the data-line signals are "locked out" from the flip-flop whenever the clock is low. Observe also in the timing diagram (Fig. 5-10A) that if the signal on the data line happens to go high-low-high-low while the clock is high, the Q output will also go high-low-high-low. In digital parlance, any change in logic level on the D line will be *clocked in* while the clock is high.

Of course, if the preset terminal is driven high at any time whatsoever, the Q output is *forced to 1*. Similarly, if the clear input is driven high at any time whatsoever, the Q output is *forced to 0*. Note that clocked D flip-flops may not always be provided with preset and clear terminals. If an application does not require asynchronous flip-flop control, the preset and clear functions may be omitted or tied to ground (not used).

COMMERCIAL IC PACKAGE

A commercial four-bit D-type latch package that is frequently encountered in logic diagrams is shown in Fig. 5-11. (It is a flip-flop package, in the strict sense of the term.) This particular design does not provide preset and clear functions. The IC package contains four latches, and four bits (a *nibble*) can be stored for one clock cycle. Each latch obeys the truth table shown in Fig. 5-12A.

Note from Fig. 5-11 that a clock input is provided for latches 1 and 2 and another clock input is provided for latches 3 and 4. In many applications, the two clock inputs are tied together. How-

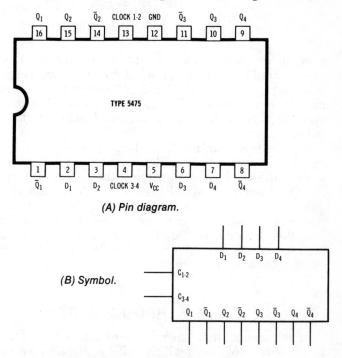

(A) Pin diagram.

(B) Symbol.

Fig. 5-11. A four-bit D-type latch IC package.

105

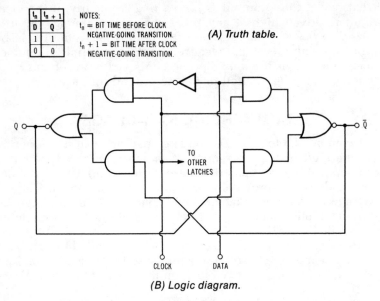

t_n	$t_n + 1$
D	Q
1	1
0	0

NOTES:
t_n = BIT TIME BEFORE CLOCK NEGATIVE-GOING TRANSITION.
$t_n + 1$ = BIT TIME AFTER CLOCK NEGATIVE-GOING TRANSITION.

(A) Truth table.

(B) Logic diagram.

Fig. 5-12. Details of D-type flip-flop.

ever, in dual-clock systems, differently phased clock signals may be applied to each pair of latches. Observe that each of the two lower AND gates in Fig. 5-12B seems to have only one input. As mentioned previously, this means that the inputs to these AND gates are tied together, and each gate functions as a buffer.

It is evident from the logic diagram in Fig. 5-12B that when the data line goes logic-high, the $\overline{Q}$ output will go logic-low as soon as the clock goes high; since the $\overline{Q}$ output goes logic-low, the Q output must simultaneously go logic-high. If the clock goes low with the data line remaining logic-high, the flip-flop does not change state. However, when the data line goes logic-low, the Q output will go logic-low as soon as the clock goes high. Since the Q output goes logic-low, the $\overline{Q}$ output must simultaneously go logic-high.

FEED-THROUGH LATCH

At this point, it is instructive to consider briefly the *feed-through* type of latch, which will be encountered on occasion in various digital logic diagrams. A feed-through latch, also

106

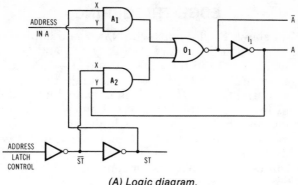

(A) Logic diagram.

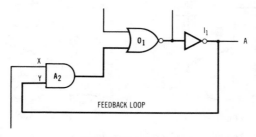

(B) Circuit operation.

Fig. 5-13.　Feed-through latch.

known as a *fall-through* latch, is depicted in Fig. 5-13. "Address" is synonymous with data in the present discussion. When the address latch-control signal is high, the address data "falls through" the latch. When the address latch-control signal is low, the address data is latched into the circuit.

Observe in Fig. 5-13 that when the address latch-control signal is high, AND gate A_1 is enabled, and AND gate A_2 is disabled. Accordingly, data "falls through" the latch. When the address latch-control signal goes low, AND gate A_1 is disabled, thereby blocking any new data inputs. At the same time, AND gate A_2 is enabled by a logic-high level at its X input, which allows the feedback state to determine the output of gate A_2. If the feedback state is low, the output of A_2 will be low. If the feedback state is high, the output of A_2 will be high. Note that there are two inversions from the output of gate A_2 (O_1 and I_1) to the feedback loop. Consequently, the feedback action maintains the level that was present on inverter I_1 when the address latch-control state went low.

EDGE TRIGGERING

Fig. 5-14 shows RS flip-flops with NAND-gate implementation. Observe that when S = 1 and R = 0, Q = 1 and $\overline{Q}$ = 0. Conversely, when S = 0 and R = 1, Q = 0 and $\overline{Q}$ = 1. Next, note the configuration in Fig. 5-15; the clock is high, the data input is high, Q = 1, and $\overline{Q}$ = 0. If the clock remains high and the data input goes low (Fig. 5-16), Q remains 1 and $\overline{Q}$ remains 0. If the clock is low and the data input goes high (Fig. 5-17), Q remains 1 and $\overline{Q}$ remains 0. Accordingly, it might seem that it would be impossible to drive Q = 0 and $\overline{Q}$ = 1. Note, however, that only steady-state conditions have been considered so far; let us observe a transient response.

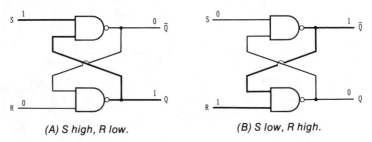

(A) S high, R low.   (B) S low, R high.

Fig. 5-14.   The two states of RS latches used in edge-triggered flip-flops.

Transient response involves propagation delays, and in this configuration, one of the propagation delays is significant. With reference to the clock-low, data-low diagram in Fig. 5-18, note that the clock input to U_3 is low and that the other two inputs to U_3 are high. Observe also that the data input is low. With these states present, it is possible to edge-trigger the configuration and thereby drive Q = 0 and $\overline{Q}$ = 1. If the clock now goes high, all three inputs of U_3 are instantaneously high. Therefore, the output of U_3 is instantaneously driven low. In turn, Q goes low and $\overline{Q}$ goes high. After a very short propagation delay (through U_3), both inputs of U_4 will be low; consequently, the data input is "locked out"; Q remains low and $\overline{Q}$ remains high, regardless of the logic level on the data input (see Fig. 5-19).

Of course, when the data input is high, Q and $\overline{Q}$ will reverse their states at the instant when the clock goes high. After a very short propagation delay, the data input is again "locked out." This is the principle of edge-trigger action. The propagation delay is approximately 10 ns; thus, there is a "window" of 10 ns during which the data input is not locked out and the flip-flop can be triggered. The advantage of edge triggering is that the

108

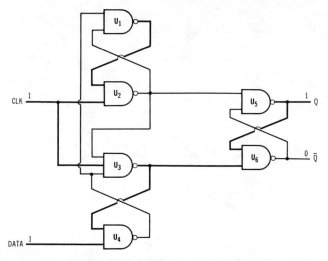

Fig. 5-15.   Clock high, data high.

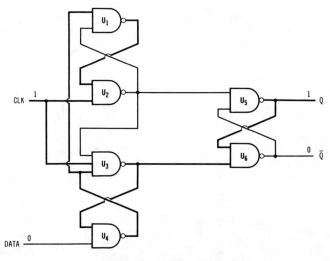

Fig. 5-16.   Clock high, data low.

109

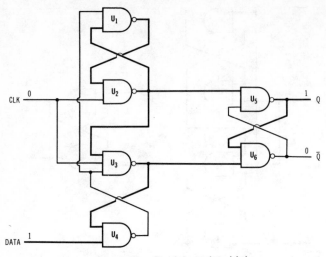

Fig. 5-17.   Clock low, data high.

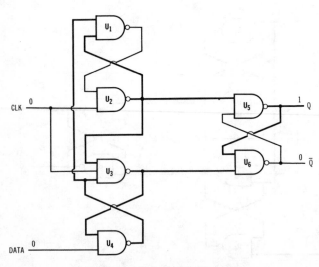

Fig. 5-18.   Clock low, data low.

110

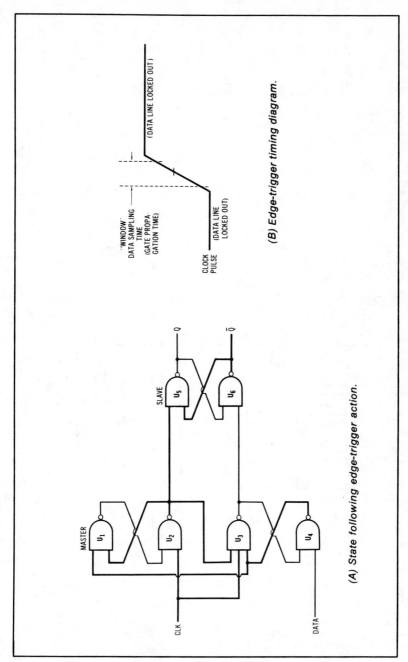

Fig. 5-19. Transient response of edge-triggered flip-flop.

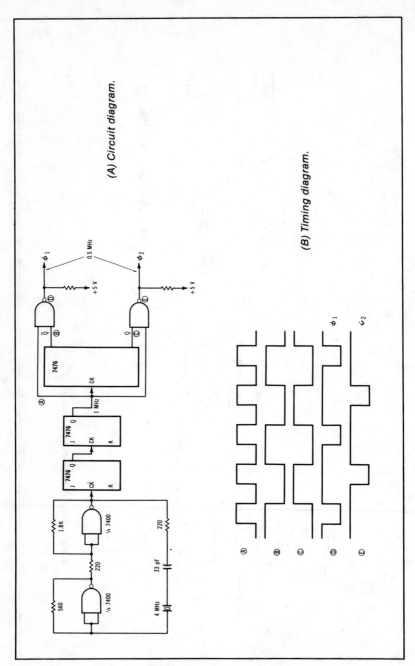

Fig. 5-20. Crystal-controlled clock oscillator.

112

data line is freed for other uses at any time except for the brief instant during which the clock goes from low to high.

CRYSTAL-CONTROLLED CLOCK

Most digital-computer systems employ a quartz-crystal clock oscialltor. An example is shown in Fig. 5-20. This is a simple two-phase 0.5 MHz configuration; it utilizes a 4-MHz quartz crystal in the positive feedback loop of two NAND gates. A frequency divider that consists of two flip-flops is followed by another flip-flop and a pair of NAND gates arranged to provide two clock output waveforms. Thus, waveforms D and E (ϕ_1 and ϕ_2) are 180° out of phase with each other.

Chapter
6

Additional Flip-Flops
and Monostables

This chapter will introduce additional types of flip-flops, and the monostable multivibrator will also be described. A discussion of digital voltage levels concludes the chapter.

EDGE-TRIGGERED D FLIP-FLOP

An edge-triggered D flip-flop configuration is shown in Fig. 6-1. This is a widely used arrangement, identical to the edge-triggered RS flip-flop described in the preceding chapter. Three RS flip-flops are used in a configuration that includes a three-input NAND gate. Whereas the individual RS flip-flops have a potential race hazard, the configuration in Fig. 6-1 eliminates any possibility of racing. An edge-triggered D flip-flop, unlike a basic RS flip-flop, has only one data input. The operation of this flip-flop is shown by the timing diagram in Fig. 6-2.

The input latches in Fig. 6-1 are interconnected so that when the positive-going edge of the clock waveform is applied, the input latches lock in complementary states. That is, one input latch always supplies a logic 1 to the output latch, and the other input latch always supplies a logic 0 to the output latch. The state in which the input latches lock depends on whether the data line is high or low when the positive-going edge of the clock pulse is applied. After the clock has gone high, both input latches are locked in their existing states, and the data line has no further control. When the clock goes low, both of the input latches supply a logic 1 to the output latch, and the data line can affect only the status of gates U_1 and U_4.

An example of a commercial dual D-type edge-triggered flip-

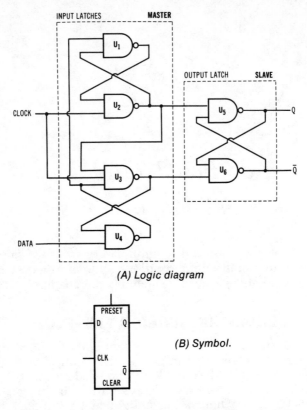

(A) Logic diagram

(B) Symbol.

Fig. 6-1. An edge-triggered master-slave D-type flip-flop.
(*Courtesy Hewlett-Packard*)

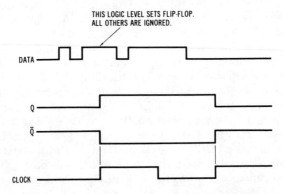

Fig. 6-2. Timing diagram for edge-triggered D-type
flip-flop. (*Courtesy Hewlett-Packard*)

flop is shown in Fig. 6-3. Positive-edge triggering is provided. Direct clear and preset inputs are included, whereby the flip-flop can be preset or cleared asynchronously and without regard to the states on the data and clock lines. Complementary (Q and $\overline{Q}$) outputs are available. Clock triggering occurs at a threshold voltage on the leading edge of the clock pulse; it is not directly related to the transition time of the leading edge. After the clock input voltage has been passed, the data input (D) is locked out. A low input to preset sets Q to the high level; a low input to clear sets Q to the low level.

An edge-triggered flip-flop should not be confused with a *ones-catching* flip-flop. A true edge-triggering flip-flop is responsive to input conditions only during the leading edge of the clock pulse. On the other hand, a ones-catching flip-flop responds to input conditions all the time that the clock is high; the data input is not locked out until the clock goes low. The clocked D flip-flop shown in Fig. 6-4 is an example of a ones-catching flip-flop.

BASIC JK FLIP-FLOP

The JK flip-flop is the most widely used design in digital logic systems; it is also the most sophisticated and versatile type of flip-flop. The basic arrangement for a JK flip-flop is shown in Fig. 6-5. It can be compared with an RS flip-flop in that it has two data inputs; however, a JK flip-flop has none of the shortcomings of an RS flip-flop. A JK flip-flop cannot have an undefined output, and its latches are not subject to a race condition. (See truth table in Fig. 6-5D.)

Observe that the configuration in Fig. 6-5 is a ones-catching arrangement, whereas the configuration in Fig. 6-6 is edge-triggered. The edge-triggered version is a master-slave circuit; it has an output that depends solely on the states of the J and K inputs at the instant of clock transition. Most edge-triggered versions are controlled by the trailing edge; that is, they are negative-edge triggered.

APPLICATION FOR JK FLIP-FLOP

An example of an application for a JK flip-flop is shown in Fig. 6-7. This is a clock circuit for a Pong™ video game. The clock frequency is established by a 14-MHz crystal oscillator in a two-inverter configuration. (The output from inverter U_2 is fed back to the input of inverter U_1, thereby sustaining oscillation.)

117

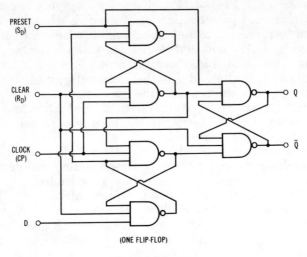

PRESET (S_D)

CLEAR (R_D)

CLOCK (CP)

D

(ONE FLIP-FLOP)

(A) Logic diagram.

t_n	t_{n+1}	
INPUT D	OUTPUT Q	OUTPUT $\bar{Q}$
0	0	1
1	1	0

(B) Truth table.

NOTES:
t_n = BIT TIME BEFORE CLOCK PULSE
t_{n+1} = BIT TIME AFTER CLOCK PULSE

TYPE 9N74
DIP (TOP VIEW)

V_{CC} 14 | $\bar{R}_{D2}$ 13 | D_2 12 | CP_2 11 | $\bar{S}_{D2}$ 10 | Q_2 9 | $\bar{Q}_2$ 8

D R_D Q   CP S_D Q
CP S_D Q̄   D R_D Q

$\bar{R}_{D1}$ 1 | D_1 2 | CP_1 3 | $\bar{S}_{D1}$ 4 | Q_1 5 | $\bar{Q}_1$ 6 | GND 7

(C) Package pinout.

Fig. 6-3. Dual D-type edge-triggered flip-flop. *(Courtesy Fairchild Camera and Instrument Corp.)*

118

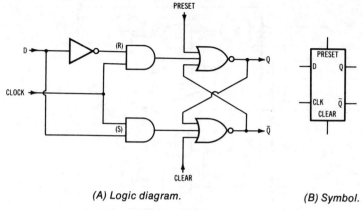

(A) Logic diagram.   *(B) Symbol.*

Fig. 6-4.   Ones-catching type of clocked D flip-flop.

The output from the crystal oscillator drives a squarer (NAND gate U₃ operating as an inverter). In turn, the shaped output from the squarer drives the clock-input terminal of a JK flip-flop. The flip-flop has an active-low clock input. It operates as a toggle (changes output state with each active clock transition) because its J and K inputs are tied together and returned to V_{CC}. The flip-flop divides by two, and a 7-MHz clock is available at its Q terminal.

OPERATION OF JK FLIP-FLOP

The waveform in Fig. 6-8 represents the clock pulse applied to a JK flip-flop. This illustration shows a mode of operation that is called *negative-edge triggering* because data can be accepted from the J and K inputs only for a brief interval prior to point 3 on the clock pulse. At time 3, the J and K inputs are disabled, and they remain disabled until the following positive-going edge rises to point 2. Although data on the J and K lines may change during the interval from 2 to 3, only the state of the data just prior to 3 will affect the following state of the slave.

With reference to Fig. 6-8, a JK flip-flop operates as follows: If one of the inputs is logic-high and the other input is logic-low, the flip-flop will be set or reset by the clock edge. If both inputs are logic-low when the clock edge occurs, the outputs remain in the same state as before clock-edge arrival. If both inputs are held logic-high, the outputs will change states when the clock edge occurs; that is, if the flip-flop was set before the clock pulse, it will reset afterward, and vice versa. This is called the *toggle* mode of flip-flop operation.

119

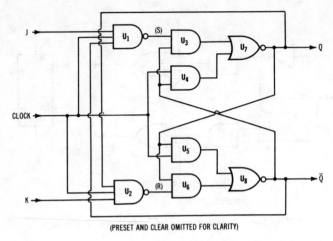

(PRESET AND CLEAR OMITTED FOR CLARITY)

(A) Logic diagram.

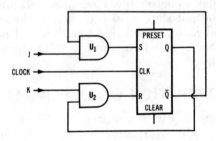

(B) Equivalent diagram.

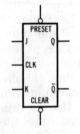

(C) Logic symbol.

INPUTS (at t_n)		OUTPUT (at t_{n+1})
J	**K**	**Q**
0	0	Q_n
0	1	0
1	0	1
1	1	$\overline{Q}_n$

(D) Truth table.

Fig. 6-5. Basic arrangement of JK flip-flop.

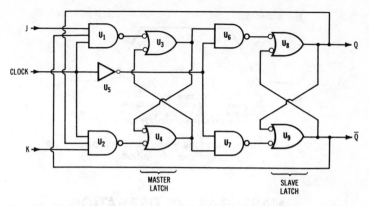

Fig. 6-6. Master-slave version of JK flip-flop. (*Courtesy Hewlett-Packard*)

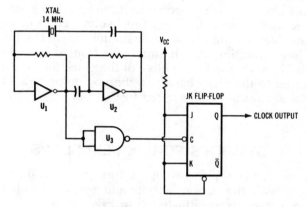

Fig. 6-7. Application for JK flip-flop in video game.

The configuration depicted in Fig. 6-5 is similar to an RS flip-flop with its outputs cross-connected back to its inputs and ANDed with the J and K inputs. The J and K output signals appear at the outputs of gates U_1 and U_2 *after* the clock signal is raised to logic-high. If the status of the J and K lines is changed at this time, these changes will be reflected at the outputs of U_1 and U_2. The cross connection of the outputs back to the input gates determines the net input conditions to (R) and (S) when J and K are both logic-high. This circumstance causes the toggling action that was noted above; an undefined or race output condition is avoided.

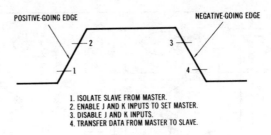

1. ISOLATE SLAVE FROM MASTER.
2. ENABLE J AND K INPUTS TO SET MASTER.
3. DISABLE J AND K INPUTS.
4. TRANSFER DATA FROM MASTER TO SLAVE.

Fig. 6-8. Timing relations along the clock pulse for a
master-slave JK flip-flop.

MASTER-SLAVE OPERATION

A master-slave implementation is shown in Fig. 6-6.
(Master-slave circuitry will also be encountered in flip-flops
other than the JK variety.) In all cases, the data is inputted into
a master latch and is subsequently transferred to a slave latch.
In a JK master-slave flip-flop, the leading edge of the clock
pulse *enables* the inputs to the first latch (master) while at the
same time isolating the master latch from the second latch
(slave). Then, on the falling edge of the clock pulse, the data is
transferred to the slave latch and thereby outputted. These sequential actions are summarized in Fig. 6-8.

SYMBOLS FOR JK FLIP-FLOPS

In general, a JK flip-flop is represented in a logic diagram by a
rectangle with the various inputs and outputs labeled. Some
common variations are illustrated in Fig. 6-9.

Fig. 6-9A shows a generalized symbol for a JK flip-flop. The
triangle at the CLK (clock) input indicates positive-edge triggering. A high level at J produces a set on the next positive-going
clock edge. A high level at K produces a reset on the next
positive-going clock edge. If both J and K are high, there will be
one toggle per positive-going clock edge.

In Fig. 6-9B, a small circle, or "bubble," is placed at the clock
input. This indicates that the flip-flop is triggered by the
negative-going edge of the clock pulse.

A standard symbol for a JK flip-flop with preset (PR) and clear
(CLR) functions is shown in Fig. 6-9C. Positive-edge triggering
is indicated. In addition, a high level at PR presets the flip-flop,
and a high level at CLR clears the flip-flop.

Fig. 6-9D represents a JK flip-flop with negative-edge triggering and low-level–activated preset and clear inputs. A low level at PR presets the flip-flop, and a low level at CLR clears the flip-flop.

In Fig. 6-9E, the absence of a triangle at the clock input indicates level triggering rather than edge triggering. Thus, a ones-catching flip-flop is represented. There is no small circle at the clock input, so this flip-flop is triggered by the high clock level. In Fig. 6-9F, there is a small circle at the clock input, so this flip-flop is triggered by the low clock level. (The clock waveforms in Fig. 6-9 are shown for purposes of explanation and are not part of the symbols.)

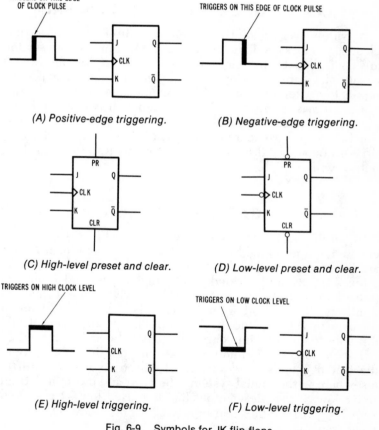

(A) Positive-edge triggering.

(B) Negative-edge triggering.

(C) High-level preset and clear.

(D) Low-level preset and clear.

(E) High-level triggering.

(F) Low-level triggering.

Fig. 6-9.   Symbols for JK flip-flops.

FLIP-FLOP FAMILY RELATIONSHIPS

Fig. 6-10 shows the relationships among the different types of flip-flops. The RS flip-flop is the primal form, which in turn was developed into the D, T (toggle), and JK versions. A T-type flip-flop was also developed from the D flip-flop. The D and T configurations may be regarded as subclasses of the JK arrangement. The JK and RS flip-flops have various minor subclasses with respect to preset and clear functions, as shown in Chart 6-1. The preset and clear functions serve to initialize the output state of the flip-flop asynchronously; that is, the flip-flop responds to these inputs at any time, and not just at clock transitions.

EXAMPLE OF JK FLIP-FLOP IC PACKAGE

A widely used dual JK master/slave flip-flop with separate clears and clocks is depicted in Fig. 6-11. Inputs to the sections are controlled by the clock pulse (Fig. 6-11C) as follows: (1) isolate slave from master, (2) enter information from J and K inputs to master, (3) disable J and K inputs, (4) transfer information from master to slave. The clear function is independent of the J and K lines and of the clock; in other words, the flip-flop can be cleared asynchronously. When the clear input (R_D) is driven low at any instant, the Q output will go low and the $\overline{Q}$ output will go high.

GATED FLIP-FLOPS

Gated flip-flops, such as the one shown in Fig. 6-12, are often encountered in digital logic circuits. In this example, AND gates are inserted in the J and K input channels. Three J inputs and three K inputs are provided. Accordingly, J_1, J_2, and J_3 must be pulsed logic-high simultaneously to activate the J channel. Similarly, K_1, K_2, and K_3 must be pulsed logic-high simultaneously to activate the K channel. This feature permits the flip-flops to be gated into or out of the data lines. For example, if J_1 is held low, the flip-flop will be unresponsive to activity on the J_2 and J_3 lines; however, this J_2-J_3 activity may be utilized elsewhere in the digital system. The clear and preset inputs are active-low; if the preset input (S_D) is pulsed low, Q goes high asynchronously; if the clear input (R_D) is pulsed low, Q goes low asynchronously.

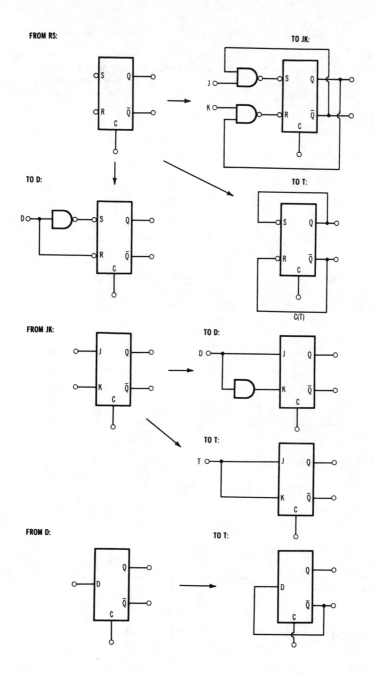

Fig. 6-10. Relationships among types of flip-flops.

Chart 6-1. Preset and Clear for JK and RS Flip-Flops

JK AND RS FLIP-FLOPS WITH PRESET OR CLEAR INPUT

Asserting the preset input or asserting the clear input overrides all other inputs. Note that the flip-flops shown below have active-low preset or clear inputs. In other words, the input is asserted by applying a logic-low level; if a logic-high level is applied, the input is unasserted.

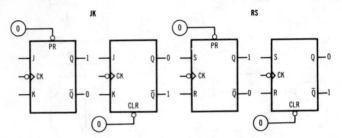

When the preset input or the clear input is unasserted, the outputs do not change until the clock pulse arrives; then, the outputs change to correspond with the input states, as in other JK or RS flip-flops.

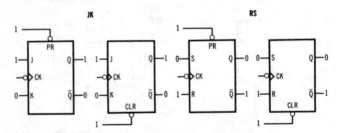

JK AND RS FLIP-FLOPS WITH PRESET AND CLEAR INPUTS

Note that the foregoing examples of flip-flops are provided with either a preset input or a clear input, but not both. The flip-flops shown below are provided with both preset and clear inputs. When the preset input is asserted (but the clear input is not asserted), the Q output goes logic-high, and all other inputs are irrelevant.

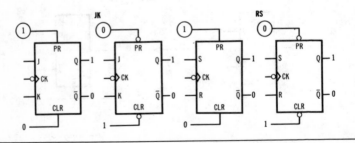

126

Chart 6-1. Preset and Clear for JK and RS
Flip-Flops (cont)

When the clear input is asserted (but the preset input is not asserted), the Q output goes logic-low, and all other inputs are irrelevant.

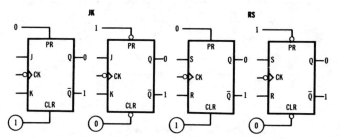

When the preset input and the clear input are both asserted, both the Q and the $\overline{Q}$ are driven logic-high but will not persist, even if both the preset input and the clear input are then unasserted at the same time (race condition occurs).

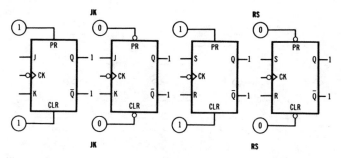

When the preset input is unasserted after the clear input has been unasserted, the Q output will remain logic-high until a clock pulse arrives.

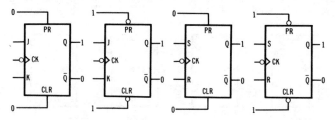

When the clear input is unasserted after the preset input has been unasserted, the Q output will remain logic-low until a clock pulse arrives.

(Continued on next page)

Chart 6-1. Preset and Clear for JK and RS Flip-Flops (cont)

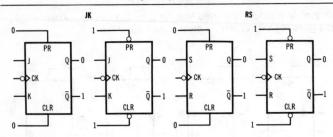

JK AND RS EDGE-TRIGGERED FLIP-FLOPS

Assuming that both the preset input and the clear input have been unasserted, if the J and K or R and S (data) inputs are in opposite logic states, the outputs change to correspond with the inputs when the trailing (negative or down-going) edge of a clock pulse arrives.

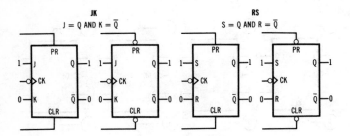

The diagrams below are the same as in the foregoing example, except that the data inputs are reversed.

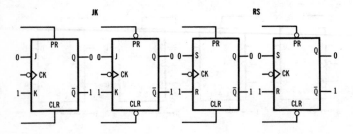

If the J and K or R and S (data) inputs are both in the logic-low state, the outputs do not change from their existing states when the down-going edge of a clock pulse arrives.

Chart 6-1. Preset and Clear for JK and RS
Flip-Flops (cont)

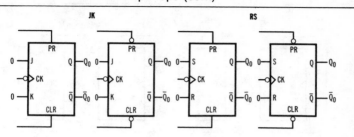

If the data inputs are both logic-high when the down-going edge of a clock pulse arrives:

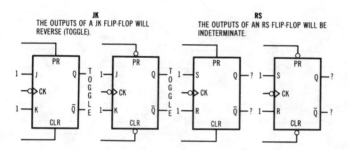

MASTER-SLAVE FLIP-FLOPS

Master-slave flip-flops recognize and store the logic states of the data inputs during and following one edge of a clock pulse, and transfer those logic states to the outputs during and following the next edge. The clock-pulse edge that causes a transfer is symbolized on the clock input line.

The logic state of a data line normally should not change during the interval between the leading edge and the trailing edge of a clock pulse. If the logic state does change during this interval, a high level probably will be stored and transferred. This is called a "ones-catching" flip-flop. Master-slave flip-flops with "data-lockout" will store only the logic level that exists during and immediately following the leading-edge clock transition.

JK AND RS PULSE-TRIGGERED (MASTER-SLAVE) FLIP-FLOPS

Assuming that the preset and clear inputs have both been unasserted, if the J and K or R and S (data) inputs were in opposite logic states during the arrival of the up-going edge of a clock pulse, the next down-going edge will transfer these stored states to corresponding outputs.

(Continued on next page)

Chart 6-1. Preset and Clear for JK and RS
Flip-Flops (cont)

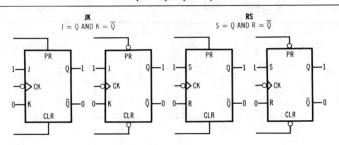

For the same conditions as above, but with the inputs reversed:

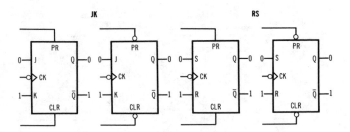

If the J and K or R and S (data) inputs were both in the logic-low state during the arrival of the up-going edge of a clock pulse, the next down-going edge will transfer no change to the outputs.

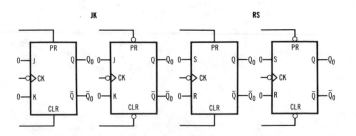

When the down-going edge of a clock pulse arrives following an up-going edge that occurred while both data inputs were logic-high:

Chart 6-1. Preset and Clear for JK and RS
Flip-Flops (cont)

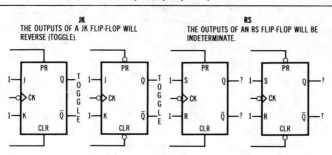

Notes:

1. Positive logic is assumed: logic-high is 1; logic-low is 0.
2. Preset (PR) is also called set (S).
3. Clock (CK) is also symbolized CP.
4. Clear (CLR) is also called reset (R).
5. A function is asserted when logic-high, unless the input line is terminated with a small circle; a small circle denotes that the function is asserted when logic-low.
6. Asserted inputs are circled, thus: ① ⓪
7. When the leading (positive, or up-going) edge of a clock pulse transfers data to the outputs, the clock input is symbolized without a small circle.

Courtesy Tektronix, Inc.

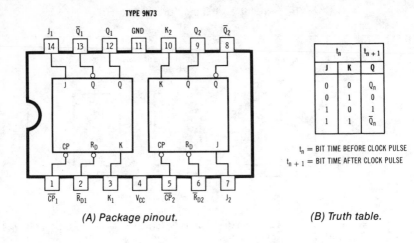

TYPE 9N73

(A) Package pinout.

t_n		t_{n+1}
J	**K**	**Q**
0	0	Q_n
0	1	0
1	0	1
1	1	$\overline{Q}_n$

t_n = BIT TIME BEFORE CLOCK PULSE
t_{n+1} = BIT TIME AFTER CLOCK PULSE

(B) Truth table.

(C) Clock waveform.

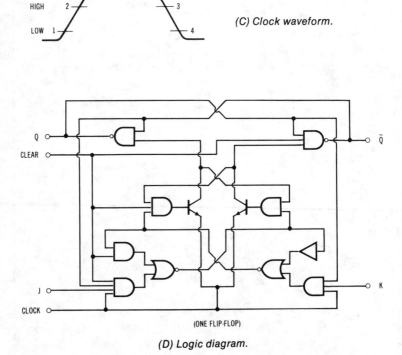

(ONE FLIP-FLOP)

(D) Logic diagram.

Fig. 6-11. Dual JK master-slave flip-flop with separate clears and clocks.
(Courtesy Fairchild Camera and Instrument Corp.)

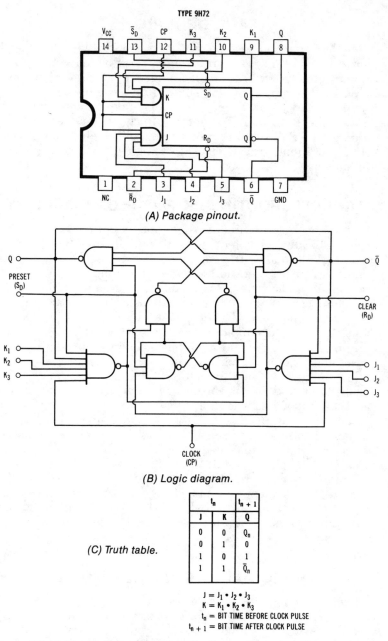

TYPE 9H72

(A) Package pinout.

(B) Logic diagram.

(C) Truth table.

t_n		t_{n+1}
J	K	Q
0	0	Q_n
0	1	0
1	0	1
1	1	$\bar{Q}_n$

$J = J_1 \cdot J_2 \cdot J_3$
$K = K_1 \cdot K_2 \cdot K_3$
$t_n = $ BIT TIME BEFORE CLOCK PULSE
$t_{n+1} = $ BIT TIME AFTER CLOCK PULSE

Fig. 6-12. Example of gated JK flip-flop configuration. (*Courtesy Fairchild Camera and Instrument Corp.*)

133

AND-OR INPUTS

Another common version of the JK master-slave flip-flop is shown in Fig. 6-13. This arrangement is provided with AND-OR inputs (actually, AND-OR-invert inputs). Inputs J_{1A} and J_{1B} must be pulsed logic-high simultaneously to activate the J channel, or inputs J_{2A} and J_{2B} must be pulsed logic-high simultaneously to activate the J channel. The same AND-OR characteristic is provided for the K channel. A preset terminal is included; this is an active-low input, and when it goes logic-low, Q goes high simultaneously.

TROUBLESHOOTING JK FLIP-FLOP CIRCUITRY

Flip-flop circuitry, like most digital circuitry, tends to fail catastrophically. As noted previously, faults may be either external or internal with respect to the IC packages. A fault generally consists of a short between interconnects, an open interconnect, a short from interconnect to ground, a short from interconnect to V_{CC}, an open bond inside an IC, an internal short to ground or to V_{CC}, or an internal fault that produces a "stuck at" trouble symptom.

In high-impedance and medium-impedance circuitry, a logic pulser and logic probe usually serve to localize a fault quickly. In low-impedance circuitry (and high-impedance circuitry with a short to ground), a current tracer is generally the most practical test instrument for fault analysis.

The following case history is presented courtesy of Hewlett-Packard. The node between U_1 and U_2 (Fig. 6-14) was found to be stuck low when checked with a logic probe, although the probe showed pulse activity at the input of U_1. Next, pin 2 was pulsed to see if the state of the node could be changed. In this example, the state could not be changed. Hence, the pulser and current tracer were utilized (Fig. 6-15), and it was found that input pin 9 on the JK flip-flop was shorted to ground. (A current tracer can indicate which end of an interconnect is shorted.)

MONOSTABLE MULTIVIBRATORS

A *monostable multivibrator*—also called a *one-shot multivibrator*, *start-stop multivibrator*, or *pulse regenerator*—is a rectangular-wave generator that has only one stable state. It produces output pulses that are independent of the input trigger pulse width. When an input trigger is applied, a monostable

TYPE 9H71

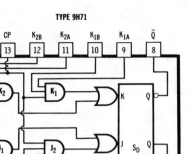

(A) Package pinout.

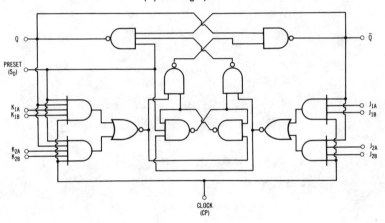

(B) Logic diagram.

	t_n		t_{n+1}
J	**K**		**Q**
0	0		Q_n
0	1		0
1	0		1
1	1		$\overline{Q}_n$

(C) Truth table.

$J = (J_{1A} \cdot J_{1B}) + (J_{2A} \cdot J_{2B})$
$K = (K_{1A} \cdot K_{1B}) + (K_{2A} \cdot K_{2B})$
t_n = BIT TIME BEFORE CLOCK PULSE
t_{n+1} = BIT TIME AFTER CLOCK PULSE

Fig. 6-13. Example of AND-OR gated JK flip-flop. (*Courtesy Fairchild Camera and Instrument Corp.*)

135

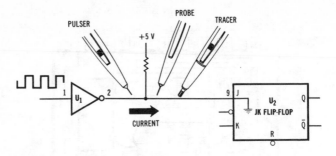

Fig. 6-14. Example of locating a short in a JK flip-flop circuit.
(*Courtesy Hewlett-Packard*)

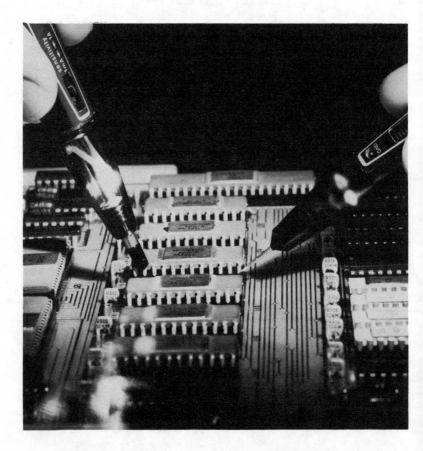

Fig. 6-15. Application of logic pulser (right) and current tracer at a
suspected node. (*Courtesy Hewlett-Packard*)

multivibrator flips to its unstable state for a period which is determined by an RC time constant. Then the monostable multivibrator returns to its original stable state. In the example of Fig. 6-16, a complementary pair of narrow output pulses is produced on each trailing edge of the input signal. The width of the output pulses depends on the values of C_X and R_X.

Fig. 6-17 shows the pinout diagram and truth table for another one-shot package. Inputs A_1 and A_2 trigger the one-shot when either or both go low when B is high. Input B triggers the one-shot when B goes high and either A_1 or A_2 is low (see truth table). An external timing capacitor may be connected between pin 10 (positive) and pin 11. With no external capacitance, the output pulse width is typically 30 ns. There is an internal 2000-ohm timing resistor, which may be used by connecting pin 9 to pin 14. For variable pulse width, an external variable resistor may be connected between pin 9 and pin 14. For accurate repeatable pulse widths, an external resistor may be connected between pin 11 and pin 14 with pin 9 open-circuited.

Elaborate types of digital-logic probes include a monostable multivibrator for operation as a "pulse stretcher." Glitches and other very narrow pulses may be encountered in digital systems. Because a very narrow pulse has much less energy than a normal pulse, it does not produce a visible indication when a simple logic probe is used. If a multivibrator is included in the probe circuitry, a clearly visible indication can be obtained when a very narrow pulse is inputted. This is because the monostable multivibrator can be triggered by a glitch and will in turn output a pulse of suitable width for energizing a LED.

Another typical application for a monostable multivibrator is shown in Fig. 6-18. This is a home-appliance application, and the unit is basically intended as a refrigerator monitor, to indicate whether or not the door has been completely closed. When light strikes phototransistor Q_1, there is no immediate response; however, after several seconds, if the light continues to strike Q_1, the alarm will start to sound. The delay interval is adjustable by means of R_8.

ONE-PULSE CIRCUIT

Older designs of digital systems utilized monostable multivibrators to a greater extent than later designs. The one-pulse circuit, shown in Fig. 6-19, is preferred by many system designers because it is a synchronized device. The one-pulse circuit employs a pair of D flip-flops and an AND gate to generate

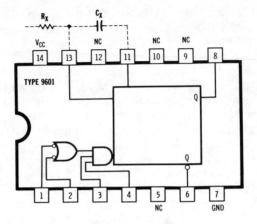

(A) Package pinout.

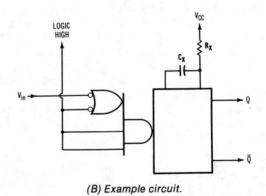

(B) Example circuit.

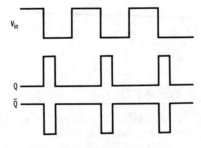

(C) Example waveforms.

Fig. 6-16. Operation of a monostable multivibrator.

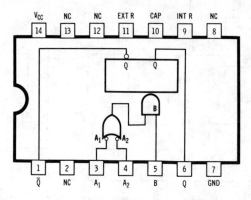

(A) Pinout diagram.

(B) Truth table.

	t_n INPUT			t_{n+1} INPUT			OUTPUT
A_1	A_2	B	A_1	A_2	B		
1	1	0	1	1	1		INHIBIT
0	X	1	0	X	0		INHIBIT
X	0	1	X	0	0		INHIBIT
0	X	0	0	X	1		ONE SHOT
X	0	0	X	0	1		ONE SHOT
1	1	1	X	0	1		ONE SHOT
1	1	1	0	X	1		ONE SHOT
X	0	0	X	1	0		INHIBIT
0	X	0	1	X	0		INHIBIT
X	0	1	1	1	1		INHIBIT
0	X	1	1	1	1		INHIBIT
1	1	0	X	0	0		INHIBIT
1	1	0	0	X	0		INHIBIT

NOTES:
1. t_n = TIME BEFORE INPUT TRANSITION.
2. t_{n+1} = TIME AFTER INPUT TRANSITION.
3. X INDICATES THAT EITHER A HIGH OR LOW
   MAY BE PRESENT.

Fig. 6-17. Monostable multivibrator package.

an output pulse that has twice the width of the clock pulse. Note that the width of the output pulse is the same whether the input of the circuit is triggered by a very narrow pulse or by a wide pulse. Thus, the circuit action is analogous to that of a one-shot multivibrator.

A one-pulse circuit is utilized chiefly as a pulse stretcher. It has the desirable feature that its output pulse width is always double the clock-pulse width, regardless of the clock frequency. This means that if the clock repetition rate changes, or if the

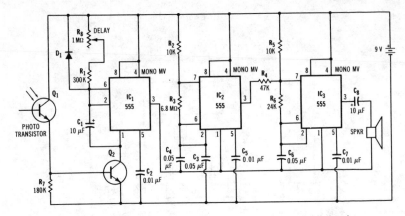

Fig. 6-18. Use of monostable multivibrator in delayed light alarm.

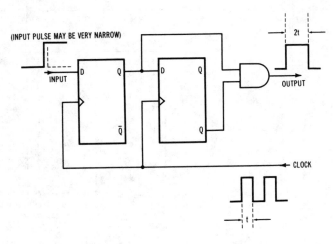

Fig. 6-19. One-pulse circuit.

digital system is operated from another clock, the system timing will be maintained. On the other hand, if a conventional one-shot multivibrator were used as a pulse stretcher, the system timing would become incorrect for any clock frequency other than that for which the one-shot was designed.

VOLTAGE VARIATION IN DIGITAL NETWORKS

A logic-high voltage value is subject to variation at different points in a digital network. A heavy load tends to reduce the high value, compared to the value for a light load. When a source drives several branch loads, it is said to have a *fan-out*. A basic buffer may function to isolate a driver circuit from the reaction of the driven circuit; it may also provide higher fan-out than the driver capability. It is customary to normalize the input and output loading parameters of TTL devices to the following values:

1 Unit TTL Load (U.L.) = 40 μA in the high state (logic 1)
= 1.6 mA in the low state (logic 0)

For example, suppose a gate has input current ratings of 1.6 mA in the low state and 40 μA in the high state. According to the definition of unit load given above, this device therefore has an input load factor of 1 U.L. (This can also be called a *fan-in* of 1 load.) Suppose another device has input current ratings of 3.2 mA in the low state and 80 μA in the high state. Since these ratings are twice the values given in the definition, this device has an input load factor of 2 U.L. Now suppose the output of a device will sink 16 mA in the low state and source 800 μA in the high state. The normalized output drive factor for this device is 16 mA ÷ 1.6 mA = 10 U.L. in the low state. The factor for the high state is 800 μA ÷ 40 μA = 20 U.L.

In theory, TTL low and high voltages are idealized at 0 volts and 5 volts, respectively. However, in practice, an input voltage in the range from 0 to 0.8 volt is within low tolerance. Low-state input voltage is designated as V_{IL}, where "IL" denotes "input low." In this context, 0.8 volt is the worst-case value. Again, in practice, an input voltage in the range from 2 to 5 volts is within high tolerance. High-state input voltage is designated as V_{IH},

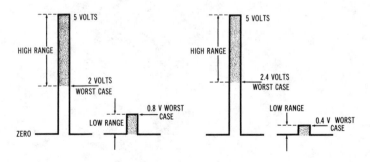

(A) Device input voltages.   (B) Device output voltages.

Fig. 6-20.   Voltage tolerances in TTL digital systems.

where "IH" denotes "input high." In this context, 2 volts is the worst-case value. Next, the output voltage may be as great as 0.4 volt in the low state; this is the worst-case value. Low-state output voltage is designated as V_{OL}, where "OL" denotes "output low." The output voltage may be as small as 2.4 volts in the high state; this is the worst-case value. High-state output voltage is designated as V_{OH}, where "OH" denotes "output high." These voltage ranges are summarized in Fig. 6-20.

In normal operation, maximum rated fan-out leaves the digital pulse voltage above its minimum rated value. However, a fault that demands additional current over the maximum rated fan-out will cause the digital pulse voltage to fall below its minimum rated value.

Chapter
7

Registers

A flip-flop is used for temporary storage of one bit of data. When a group of flip-flops is used for temporary data storage, the arrangement is called a *register*. There are various types of registers with which the digital technician must be familiar. An elementary four-bit latch-type register is shown in Fig. 7-1. It consists of four D-type flip-flops with wire-AND implementation. The capacity of the register is four bits, or one nibble. Inputs D_0, D_1, D_2, and D_3 are for the incoming data. If the set inputs ($\overline{S}_0, \overline{S}_1, \overline{S}_2$, and $\overline{S}_3$) are active (logic-low), the applied data bits will enter the register when the enable input ($\overline{E}$) is active (logic-low). Any previously stored data is automatically erased, and the new data is available at outputs Q_0, Q_1, Q_2, and Q_3. Note that the four flip-flops are independent of each other and can be used separately if desired. Input $\overline{MR}$ (active low) is a master reset input.

BASIC SHIFT REGISTERS

Shift registers are often encountered in logic circuits. A shift register is more versatile than a latch-type register; it serves as a temporary memory for storage of data, and it can also multiply by 2 or divide by 2. Many shift registers can convert parallel data into serial data, or vice versa. *Parallel* means that the data word, such as 1101, is loaded or unloaded in a single operation (all bits transferred simultaneously), whereas *serial* means that the bits in the data word are inputted or outputted sequentially.

An example of a basic serial shift register is shown in Fig. 7-2. This register employs clocked D flip-flops and has a capacity of four bits. The Q output from one flip-flop is fed (clocked) into the D input of the next flip-flop. All of the flip-flops are clocked

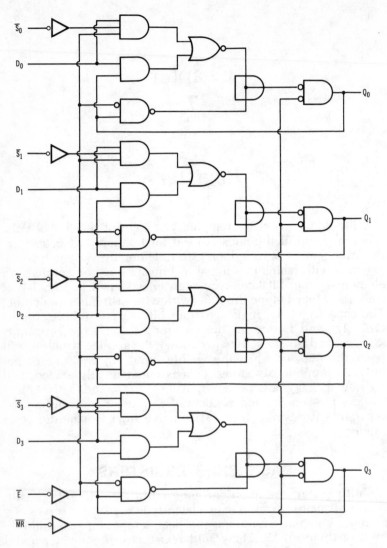

Fig. 7-1. Logic diagram of elementary 4-bit latch-type register.
(*Courtesy Fairchild Camera and Instrument Corp.*)

from the same source; if the flip-flops are cleared, they are all cleared simultaneously. In practice, the flip-flops are all cleared (reset) initially (Fig. 7-3). A logic 1 is clocked into the first flip-flop, where it is stored for one clock cycle. When the next clock pulse arrives, the 1 at the Q output of FF1 is clocked into FF2,

144

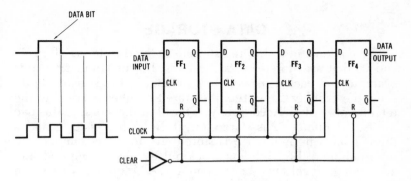

Fig. 7-2. Logic diagram of a basic shift register.

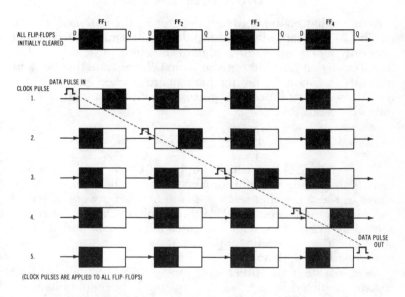

Fig. 7-3. Shift of a data pulse into, through, and out of a register.

and the 1 is now stored in FF2. (Note that when the 1 is clocked out of FF1 a new data bit can be clocked into FF1.) The 1 that is stored in FF2 is clocked into FF3 and then clocked into FF4. From FF4, the 1 bit is unloaded on the next clock pulse, or we say that the register overflows. It is evident that the 1 bit which was clocked into FF1 has been *stepped,* or *shifted,* from one flip-flop to the next in synchronism with the clock signal. This 1 bit is called a *data bit.* Of course, 0 is also a data bit.

145

DATA STORAGE

The contents of a shift register are defined in terms of the 1s and 0s that are stored in its flip-flops. Let us stipulate in this example that each 1 bit which is present in the register represents the number 1 and that the stored number is equal to the total number of 1s. Accordingly, if the shift register contains two 1 bits, the number 2 is stored. Or, if the shift register contains three 1 bits, the number 3 is stored. To store the number 4 in the example of Fig. 7-2, four clock pulses must be applied, and a logic-high level must be present at the D input of FF1 when the leading edge of each clock pulse occurs.

DATA SHIFTING

A somewhat elaborated shift-register configuration is shown in Fig. 7-4. Observe that the flip-flops will sample serial input data only when the leading edge of the clock pulse occurs. Let us suppose that the shift register is first cleared and that the four data bits shown in the timing diagram are clocked into the register. At the end of four clock pulses, the flip-flops will have the following states:

$$FF1 = 0 \quad FF2 = 1 \quad FF3 = 1 \quad FF4 = 1$$

These states can be written simply as 0111, and this term represents the number 3 in accordance with the definition that we chose in the preceding section. If the number 2 were stored, it would be represented as 0011; the number 4 would be represented by 1111. Thus, with four cascaded flip-flops and our chosen definition of stored numbers, it is possible to store five numbers: 0, 1, 2, 3, and 4.

Next, consider the *functional modes* provided by the shift register depicted in Fig. 7-4. As explained above, the bits were loaded sequentially (one after the other), and they were shifted from one flip-flop to the next. Therefore, this functional mode is called *serial loading* of data, and the configuration is called a *four-bit serially loaded shift register*. The other basic mode of loading a shift register is called *parallel leading*. In this functional mode, a separate line is connected to the preset input of each flip-flop, as seen in Fig. 7-4. All of the data bits may be loaded into the shift register simultaneously by setting the appropriate flip-flops logic-high via the preset inputs. It follows from previous discussion that this is an asynchronous functional

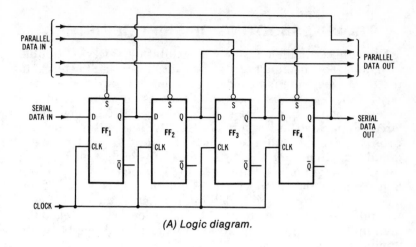

(A) Logic diagram.

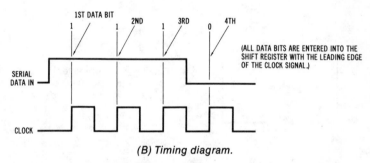

(B) Timing diagram.

Fig. 7-4. Shifting of data into and out of a register. (*Courtesy Hewlett-Packard*)

mode for loading the shift register. The data on the four preset inputs is called *parallel data*.

Observe next that just as data bits can be loaded into the shift register in either a parallel mode or a serial mode the data bits can also be read out either one by one or simultaneously. *Parallel readout* is accomplished by simultaneously sampling the data at the outputs of each of the flip-flops in Fig. 7-4. *Serial readout* is accomplished by shifting (clocking) the stored data bits through the flip-flops and sampling them progressively (sequentially) at the output of FF4. Note also that if the data is loaded serially and read out in parallel, the shift register may be said to be operating as a *serial-to-parallel converter*. On the other hand, if the data is loaded in parallel and shifted out serially, the register may be said to be operating as a *parallel-to-serial converter*.

TROUBLESHOOTING OF SHIFT REGISTERS

Sometimes a shift register will exhibit a "stuck at" trouble symptom. This malfunction can be caused by a fault external to the IC or by an internal fault. To distinguish between these two possibilities, a current tracer is useful. As an example, in one case outputs A, B, C, and D (Fig. 7-5) were stuck low; other circuits appeared to be normal. Tests with a pulser and logic probe revealed that the pins that were stuck low were grounded. (To make these tests, each pin is probed and pulsed; if a pin is ungrounded, the state will be changed by the pulse.) Other pins on the IC appeared to be normal when tested. Then a current tracer was used to determine whether the ground fault was internal or external to the IC. The current tracer tracked the fault up to the terminals of the IC; this result indicated that the ground fault was internal to the IC.

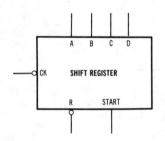

Fig. 7-5. Shift-register inputs and outputs. (*Courtesy Hewlett-Packard*)

BROADSIDE REGISTER

Broadside registers, such as the one shown in Fig. 7-6, are widely used in microprocessor systems for temporary data storage. The broadside register is a controlled type of register. Data may be loaded serially into the register when the SHL line is logic-high. Data also may be loaded in parallel (broadside). To broadside-load the register, the load line is driven logic-high. If both the load and SHL lines are held logic low, each flip-flop remains in the same state after every clock pulse; the contents of the register recirculate indefinitely (the data is "frozen"). When the load line is driven logic-low and the SHL line is driven logic-high, the configuration operates as a left-shift register. The contents of the register may be erased at any time by driving the CLR line logic-high. Stored data is outputted from the Q terminals of the flip-flops.

148

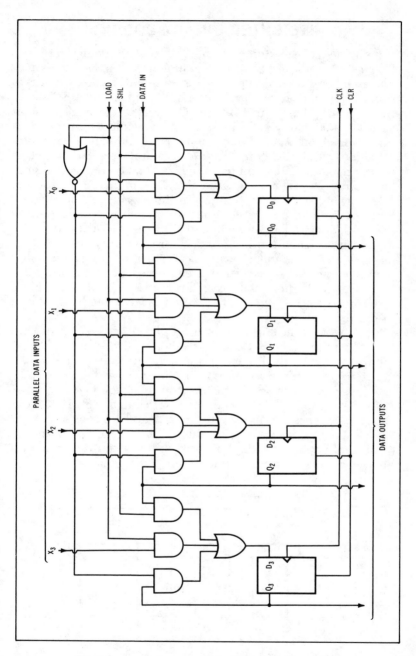

Fig. 7-6. Logic diagram of a typical broadside register.

149

WEIGHTED BINARY CODING

Four bits can be stored in the shift-register configuration in Fig. 7-4. According to the rule assumed earlier in the chapter (the number stored is equal to the number of ones stored), to store a number larger than 4, the shift register would have to contain more than four flip-flops. Obviously, if a large number such as 1,000,000 were to be stored, this would be an impractical arrangement. Accordingly, another method of number representation, called *weighted coding*, is employed. In weighted-coding notation, a data bit has a *place value* as well as a high or low value. In other words, weighted coding assigns different values (weights) to each flip-flop in the register. As an illustration, if a 5-bit shift register is utilized, the flip-flop on the right (Fig. 7-7) is assigned a weight of 1. In turn, the next flip-flop to the left is assigned a weight of 2; the remaining three flip-flops are weighted 4, 8, and 16, as indicated. Accordingly, when any one of the flip-flops is set, it then represents the weight assigned to it. If more than one flip-flop is set, the individual values (weights) are totaled to determine what number is stored in the register.

Suppose that the flip-flop on the right in Fig. 7-7 is set, and that the other four flip-flops are cleared (reset); the stored number is evidently 1. Next, if the register contents are 00101, the stored number is 4 plus 1, or 5. Observe that the zeros indicate *place values*; mathematicians refer to a zero place value as an "empty column." Thus, if the register contents are 10000, the stored number is 16, because the 8, 4, 2, and 1 places have zero values. The 16 place has a 1 value with a 16 weight, or a decimal number value of 16. A five-bit shift register can store any number up to 31; if we include 0, the register can store 32 dif-

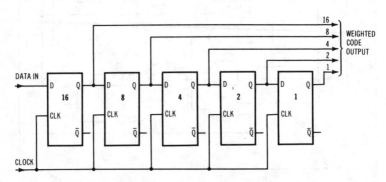

Fig. 7-7. Example of weighted coding in a five-bit shift register.

ferent numbers in weighted code. Recall that if weighted coding is not employed, this register could store only six numbers (including 0).

The foregoing weighted code is widely used and is of basic importance in digital technology. It is often called the *8421 binary code.* One version of the 8421 binary code utilizes four bits to represent decimal numbers 0 through 9 and is termed the *binary coded decimal,* or *bcd, code.* The bcd code finds extensive application in digital calculators, for example. It is apparent that the 8421 binary code is a comparatively simple code; it also provides considerable versatility. The 8421 code can be used to store very large numbers in a comparatively small number of flip-flops. Note, however, that although the 8421 code is the most widely used code, other codes are also encountered in analysis of digital logic circuits.

Observe that to store a number greater than 31, the shift register depicted in Fig. 7-7 is simply elaborated by one more flip-flop, and the weight of this new flip-flop is double that of its predecessor, or 32. It can be seen that register capacity increases rapidly, because the weight of each succeeding flip-flop is doubled—64, 128, 256, 512, 1024, and so on. A 10-bit shift register can store 1024 different numbers. Stored numbers have a *most significant bit (msb)* and a *least significant bit (lsb).* The logic 1 or 0 which is stored in the flip-flop with the smallest weight is the least significant bit. The bit which is stored in the flip-flop with the highest weight is the most significant bit. The terms lsb and msb may be used to identify the orientation of bits in any bit stream. For example:

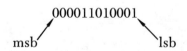

FIVE-BIT SHIFT-REGISTER IC PACKAGE

Observe the arrangement of the 5-bit shift-register IC in Fig. 7-8. This configuration is contained in a 16-pin package; it provides serial or parallel loading and serial or parallel readout (Fig. 7-8B). The shift register can be asynchronously cleared for all outputs at 0. Parallel loading occurs by way of the preset inputs of the flip-flops, which are NANDed to the data inputs and the PL (parallel-load enable) input (Fig. 7-8C). Serial-to-parallel and parallel-to-serial conversion is provided. The clear input is active-low because double inversion occurs prior to the clear inputs of the flip-flops. When double inversion takes place, the

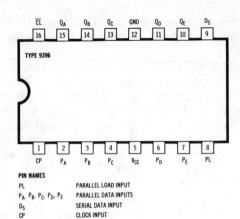

(A) Package pinout.

PIN NAMES

PL	PARALLEL LOAD INPUT
P_A, P_B, P_C, P_D, P_E	PARALLEL DATA INPUTS
D_S	SERIAL DATA INPUT
CP	CLOCK INPUT
$\overline{CL}$	CLEAR INPUT
Q_A, Q_B, Q_C, Q_D, Q_E	PARALLEL DATA OUTPUTS

PRESET COMMON	BIT	CLEAR	SERIAL INPUT	CLOCK	OUTPUT	
0	X	0	X	X	0	CLEAR ALL OUTPUTS TO 0.
1	1	1	X	X	1	PRESET OUTPUTS TO 1
1	0	1	X	X	0	INPUT BIT CONFIGURATION.
0	X	1	1	ENABLE	1	SERIAL INPUT SHIFT RIGHT.
0	X	1	0	ENABLE	0	SERIAL-TO-PARALLEL CONVERSION.

(B) Truth table.

NOTE: X = "DON'T CARE."

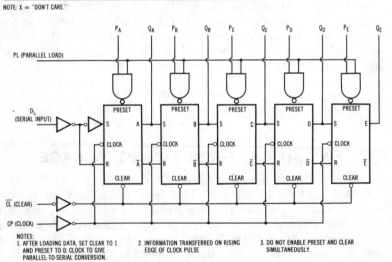

NOTES:
1. AFTER LOADING DATA, SET CLEAR TO 1 AND PRESET TO 0; CLOCK TO GIVE PARALLEL-TO-SERIAL CONVERSION.
2. INFORMATION TRANSFERRED ON RISING EDGE OF CLOCK PULSE
3. DO NOT ENABLE PRESET AND CLEAR SIMULTANEOUSLY.

(C) Logic diagram.

Fig. 7-8. Example of a five-bit shift-register IC. (*Courtesy Fairchild Camera and Instrument Corp.*)

output has the same form as the input. Use of double inversion is advantageous in some design situations to introduce sufficient propagation delay to avoid a race problem.

EIGHT-BIT SHIFT-REGISTER IC PACKAGE

The eight-bit shift-register IC package in Fig. 7-9 provides serial loading and serial readout only. The data input is ANDed, whereby serial loading can be enabled or disabled at will. Note that the truth table is compiled in terms of t_n and t_{n+8}. This is the most practical format, because a bit that is applied to the data input becomes inaccessible until after 8 clock pulses. No provision is made for clearing or presetting the shift register. All flip-flops are cleared to logic-low, if necessary, by clocking out the contents of the register. The Q output must be initially low, and the $\overline{Q}$ output must be initially high, because a preset control line is not provided.

EIGHT-BIT BIDIRECTIONAL
SHIFT-REGISTER IC PACKAGE

An eight-bit bidirectional shift-register arrangement is shown in Fig. 7-10. A 24-pin IC package is utilized. This device shifts the entered data to the right if mode-control inputs S_0 and S_1 are held logic-high and logic-low, respectively (Fig. 7-10C). Conversely, if S_0 is held low and S_1 is held high, the entered data will be shifted to the left. Note, however, that if serial data are to be entered and shifted left, the L data input is used; on the other hand, if serial data are to be entered and shifted right, the R data input is employed. Parallel data may be entered via P_A through P_H and shifted left or right, as commanded. In any mode of operation, data may be read out at all times via Q_A through Q_H. Shifted data will eventually clock out of the Q_H terminal or the Q_A terminal, depending on the direction of shift. If both S_1 and S_0 are held low, the result is to "stop the clock." Data that has been loaded in parallel, for example, can be retained until needed; the clock may then be started by driving either S_1 or S_0 high, whereupon the stored data will be clocked out to the left or to the right.

As shown in Fig. 7-11, data flow in various logic diagrams may involve serial input and serial output, serial input and parallel output, parallel input and serial output, or parallel input and parallel output.

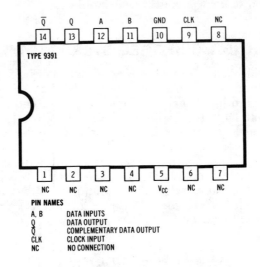

(A) Package pinout.

t_n		t_{n+8}
A	**B**	**Q**
0	0	0
0	1	0
1	0	0
1	1	1

NOTES:
t_n = BIT TIME BEFORE CLOCK PULSE
t_{n+8} = BIT TIME AFTER EIGHT CLOCK PULSES

(B) Truth table.

Fig. 7-9. Example of an

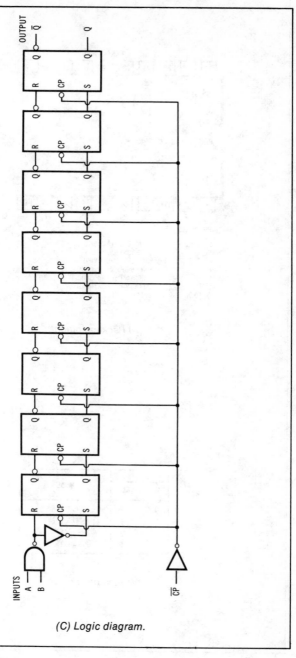

(C) Logic diagram.

eight-bit shift-register IC.

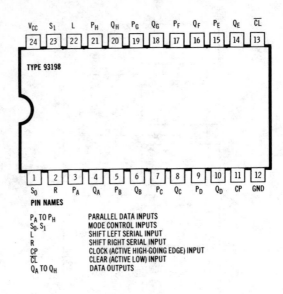

(A) Package pinout.

INPUTS		MODE
S_1	S_0	
0	0	INHIBIT CLOCK
0	1	SHIFT RIGHT
1	0	SHIFT LEFT
1	1	PARALLEL LOAD

(C) Mode control states.

Fig. 7-10. An eight-bit bidirectional shift register.

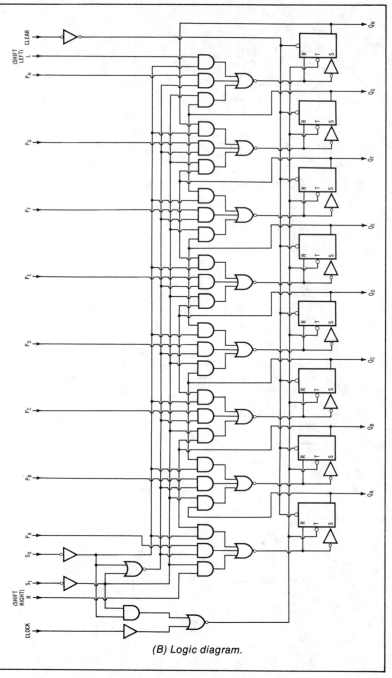

(B) Logic diagram.

(Courtesy Fairchild Camera and Instrument Corp.)

157

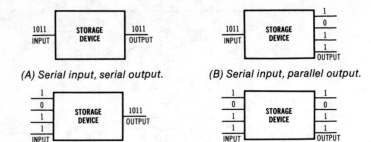

(A) Serial input, serial output.   (B) Serial input, parallel output.

(C) Parallel input, serial output.   (D) Parallel input, parallel output.

Fig. 7-11. Data-flow modes in arithmetic units, registers, memories.

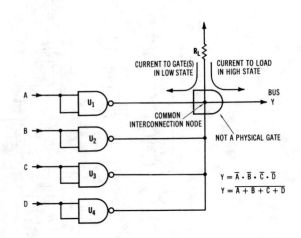

$$Y = \overline{A} \cdot \overline{B} \cdot \overline{C} \cdot \overline{D}$$
$$Y = \overline{A + B + C + D}$$

(A) Connection of gates.

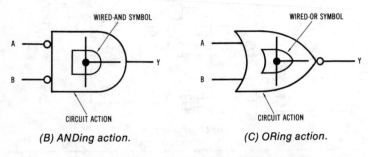

(B) ANDing action.   (C) ORing action.

Fig. 7-12. Example of wire-AND (wire-OR) operation.

BASIC TROUBLESHOOTING PROCEDURES

Fig. 7-12 shows a configuration called an *implied*-AND, *dot*-AND, *wire*-AND, or *wire*-OR connection. It can be regarded as providing ANDing action or ORing action in accordance with the equations in Fig. 7-12A. When operated with respect to ANDing action, the circuit is termed a wire-AND connection; when operated with respect to ORing action, the circuit is termed a wire-OR connection. The configuration is the same in either case. Note that in digital-logic diagrams, pull-up resistor R_L is generally omitted; the AND-symbol outline with an interconnection-node dot indicates that a pull-up resistor is used in the physical network.

This type of configuration holds the potential for a "tough dog" troubleshooting situation. The open collectors of the NAND gates are connected to a common external load resistor (Fig. 7-13). If any one of the gates goes low, it controls, or "sticks,"

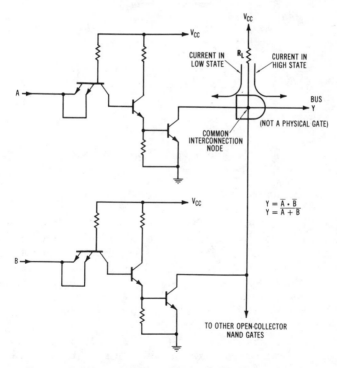

Fig. 7-13. Example of NAND gates in wire-AND application.
(*Courtesy Hewlett-Packard*).

159

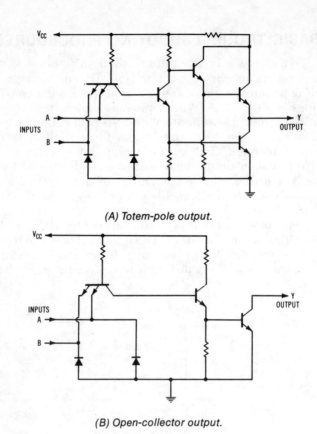

(A) Totem-pole output.

(B) Open-collector output.

Fig. 7-14. Two basic NAND-gate arrangements.

the bus, because an open-collector gate cannot supply any current in its high logic state. Therefore, if any NAND gate in Fig. 7-12 or Fig. 7-13 becomes "stuck low," all of the other gates also appear to be stuck low. Therefore, a current-tracing probe is used to troubleshoot this type of circuitry. If all of the NAND gates are driven high, only the stuck-low gate will draw current from R_L. Therefore, a current tracer can lead the technician to the defective gate.

Another potential "tough dog" situation can arise due to "cockpit error" (a wrong assumption on the part of the technician). Fig. 7-14 shows two basic forms of NAND gates; one arrangement employs the totem-pole output circuit, while the other employs the open-collector output circuit. Both arrangements are commonly represented by the same symbol in

160

digital-logic diagrams—the conventional NAND-gate symbol may represent either arrangement. Beginners tend to assume that the conventional NAND-gate symbol always implies a totem-pole output circuit. This assumption can lead to tough-dog problems and great confusion in reading digital-logic diagrams because the two arrangements are not interchangeable. A network designed for use with open-collector output will not work with totem-pole output, and vice versa. Distinction between the two basic arrangements can be made only on the basis of context of signal flow; that is, one arrangement will "make sense," and the other arrangement will "make non-sense."

Chapter
8

Counters

A counter is a circuit that sequentially counts input pulses. One basic type of counter outputs a pulse each time a predetermined number of pulses is applied to the input. A counter is sometimes called an *accumulator*. Basically, a counter is an arrangement of flip-flops containing a binary word that increases in value by one each time a pulse is inputted. Note that a counter may also be called a *frequency divider*, inasmuch as successive counter stages divide the number of input pulses by two.

Digital counters are extensively utilized in digital systems. A digital counter element is often termed a *binary counter* or *binary scaler*. It produces one output pulse for every two input pulses. Thus, a toggle flip-flop is also called a binary scaler. These devices are used wherever it is necessary to monitor the number of occurrences of a specified event, such as the application of a clock pulse or a data pulse. A symbolic convention for a four-bit binary counter is shown in Fig. 8-1A. The counter inputs a clock signal and outputs a binary-coded count on four output signal lines (coded count output). Fig. 8-1B shows the input and output waveforms. (These are typical of actual waveshapes; the distortions are within rated tolerances.)

The flip-flops and gates used in a binary counter are interconnected to advance the binary coded count output by one for each clock pulse that is applied. The sequence of the counter outputs is as follows:

(0)	0000	(4)	0100
(1)	0001	(5)	0101
(2)	0010	(6)	0110
(3)	0011	(7)	0111

(8)	1000	(15)	1111
(9)	1001	(16)	0000
(10)	1010	(17)	0001
(11)	1011	(18)	0010
(12)	1100	and so on . . .	
(13)	1101		
(14)	1110		

This type of counter employs four flip-flops—one for each output line. Note that the four flip-flops in Fig. 8-2 are interconnected directly and via AND gates. Although the AND gates could be (and sometimes are) omitted, gates provide faster action (less propagation delay). In other words, if the gates are not used, *ripple carry* then occurs, which delays outputting of the final count. Observe that the clock is connected in parallel to all flip-flops in the counter so that their output states change simultaneously. This arrangement is called a *synchronous counter*. In

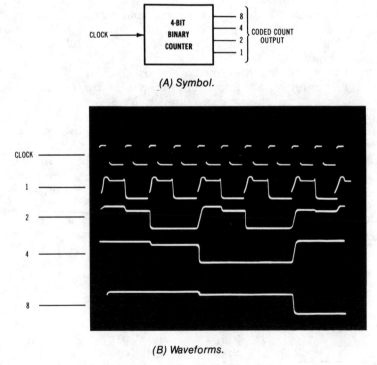

(A) Symbol.

(B) Waveforms.

Fig. 8-1. Four-bit binary counter.

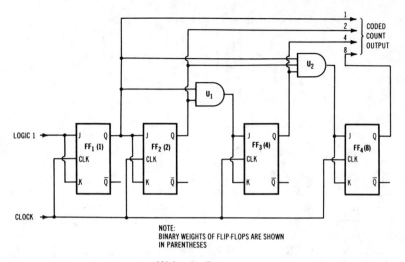

(A) Logic diagram.

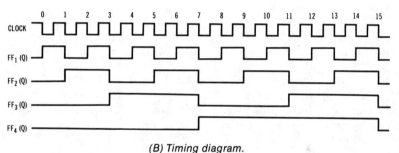

(B) Timing diagram.

Fig. 8-2. Operation of a four-bit binary counter.

this example, JK flip-flops are used in a toggle circuit; FF_1 is caused to toggle by holding both the J and K inputs logic-high.

As seen in the timing diagram (Fig. 8-2B), the FF_1 output is logic 1 just before FF_2 changes state. Since the logic 1 from FF_1 is present on alternate clock pulses only, FF_2 is toggled only on alternate clock pulses. The Q output from FF_1 has half the clock repetition rate, and the Q output from FF_2 has half the FF_1 repetition rate. To toggle FF_3, it is necessary for both FF_1 and FF_2 to be logic-high because the outputs of FF_1 and FF_2 are ANDed together to provide the pulse applied to the J and K inputs of FF_3.

Observe next the ripple-carry configuration in Fig. 8-3; suppose that the count is 0111 (note the weights assigned to the flip-flops as indicated by the numbers in parentheses). When

165

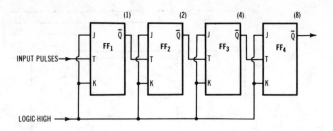

Fig. 8-3. Four-bit ripple-carry binary counter.

one more clock pulse is then applied, FF_1 changes state to 0, followed by FF_2, then followed by FF_3; finally FF_4 changes state to 1. This is a *ripple carry* process. By way of contrast, if the count in Fig. 8-2 is 0111 and one more clock pulse is then applied, FF_1 changes state to 0, and FF_4 changes state to 1 because the J and K inputs of FF_4 have been made 1 via gate U_2. Ripple carry through FF_2 and FF_3 is thus eliminated, and circuit response is faster.

The arrangement in Fig. 8-3 is also called a *mod-16* (*modulo-16* or *modulus-16*) counter because it sequences through 16 states. It may also be termed a $2 \times 2 \times 2 \times 2$ counter, because each flip-flop sequences through two states. If 20 flip-flops are connected in this manner and an input pulse is applied to FF_1 each second, approximately a minute is required for an output pulse to appear at FF_6; approximately an hour is required for an output pulse to appear at FF_{12}; approximately 12 days are required for an output pulse to appear at FF_{20}.

APPLICATION IN HOME APPLIANCE

A home-appliance application of a 14-stage ripple counter (IC_1) is shown in Fig. 8-4. This is a configuration for a door-closure monitor; it provides a delayed output ranging from 1 to 2 minutes. If a garage door, for example, has not been closed after the delay interval, the output from the counter will automatically actuate the relay and thereby close the door.

RING COUNTERS

It is apparent that shift registers and counters are similar in various aspects. By adding some gates to a shift register and thereby modifying the data path through the configuration, a counter function can be obtained. *Ring counters* are in wide use, and a ring counter is basically a circulating shift register, as

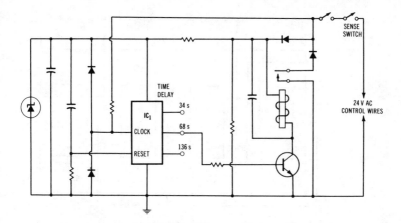

Fig. 8-4.   Counter application in a door-closure monitor.

can be seen in Fig. 8-5. In other words, it is a shift register with
its output connected back into its input. Observe that the preload line is connected to the set input of FF_1 and to the reset
inputs of FF_2, FF_3, and FF_4. The counter can be *preloaded* by
pulsing the preload input logic-low. Then, the clock pulses will
circulate the logic 1 bit in FF_1 to FF_2, to FF_3, to FF_4, and then
back to FF_1 in accordance with the truth table. (This
configuration is a mod-4 counter because it sequences through
four states.) Ring counters are used in all scanner-monitor
radios.

Operation of Scanner-Monitor Radio

A ring-counter configuration for a four-channel scanner is shown
in Fig. 8-6. Here, ring-counter output is provided by the Q and
$\overline{Q}$ outputs of two flip-flops connected to four AND gates. Observe
that the clock is provided with an inhibit input. This is a turn-
off function whereby the clock can be started or stopped as required. For example, suppose that AND gate 3 is outputting a
pulse; this pulse causes activity on channel 3 to be sampled. In
the event that the channel is active, a high level appears on the
inhibit line, and the clock is stopped. After activity on channel 3
terminates (carrier ceases), a low level appears on the inhibit
line, and the clock starts again. The clock has a repetition rate of
several hertz.

With reference to Fig. 8-6, during the time that a gate has a
logic-high output, it biases a switching diode in the following
oscillator circuit, thereby closing a quartz-crystal circuit in the
local-oscillator section. In turn, the oscillator operates at the

167

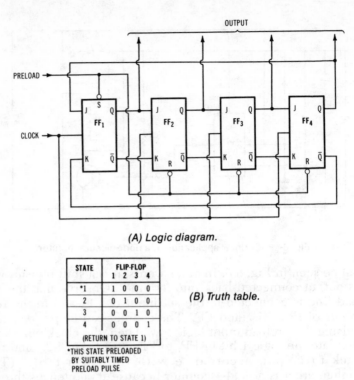

(A) Logic diagram.

STATE	FLIP-FLOP
	1	2	3	4
*1	1	0	0	0
2	0	1	0	0
3	0	0	1	0
4	0	0	0	1
(RETURN TO STATE 1)

*THIS STATE PRELOADED
BY SUITABLY TIMED
PRELOAD PULSE

(B) Truth table.

Fig. 8-5. A ring counter with preload function.

particular crystal frequency as long as the gate has a logic-high output. Then, when the gate goes logic-low, the particular crystal is switched out of the oscillator circuit, because its associated switching diode is then reverse-biased.

A timing diagram for the individual gates in Fig. 8-6 is shown in Fig. 8-7. Note that Gate 1 outputs a pulse when the $\overline{Q}$ outputs of FF_1 and FF_2 are logic-high. Next, Gate 1 turns off; Gate 2 outputs a pulse because the Q output of FF_1 is logic-high and the $\overline{Q}$ output of FF_2 is also logic-high. Then, both Gate 1 and Gate 2 are turned off; Gate 3 outputs a pulse because the $\overline{Q}$ output of FF_1 is logic-high and the Q output of FF_2 is also logic-high. Gates 1, 2, and 3 are then turned off; Gate 4 outputs a pulse because the Q output of FF_1 goes logic-high and the Q output of FF_2 also goes logic-high. The cycle then repeats.

Special States

A practical advantage of ring counters is that, unlike other types of counters, they do not require decoding. Any output line

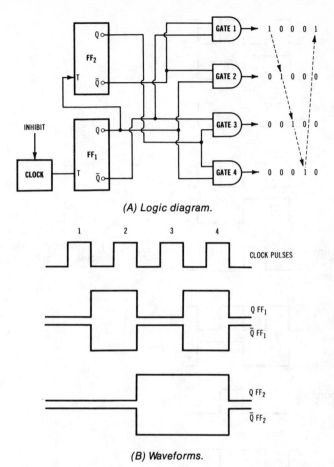

(A) Logic diagram.

(B) Waveforms.

Fig. 8-6. Ring counter for four-channel scanner-monitor radio.

from a ring counter can be connected directly to the circuit or device that it is to activate. On the other hand, ring counters have certain limitations. In the first analysis, a ring counter does not make efficient use of its flip-flops. A four-bit ring counter can generate only four unique states, whereas a four-bit binary-coded counter can generate 16 unique states. In general terms, a ring counter has N unique states, whereas a binary-coded counter has 2^N unique states, where N is the number of flip-flops in the counter. Furthermore, if a ring counter is set incorrectly due to a noise pulse or a malfunction, this erroneous ("special") state will continue to circulate indefinitely.

The possibility of erroneous sequencing in a ring counter can

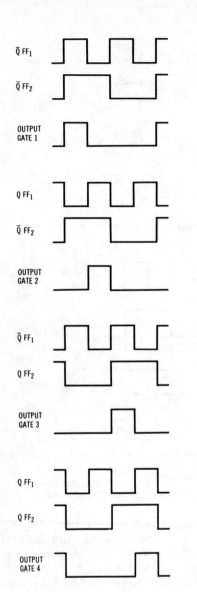

Fig. 8-7. Timing diagrams for gates in Fig. 8-6.

be avoided by employing a configuration such as the one shown in Fig. 8-8. In this implementation, any incorrect bits that start to circulate will merely be shifted through the flip-flops until all have 0 bits in them (except possibly the last flip-flop). Up to this time, the J input of the first flip-flop will be held low, and the K input will be held high. Then, after the flip-flops go to the 0001

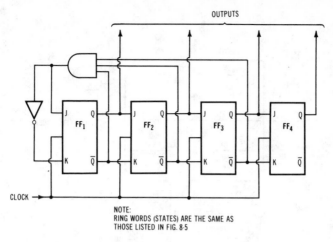

Fig. 8-8. Self-correcting/self-preloading ring counter.

state, the inputs to the first flip-flop reverse—J goes high and K goes low. The next clock pulse then sets FF_1 and resets FF_4, thereby restarting the normal sequence.

Observe that this counter configuration is *self-preloading*. When the flip-flops go the 0001 state, the $\overline{Q}$ outputs of FF_1, FF_2, and FF_3 are all logic-high, and in turn the AND gate outputs a pulse which is routed directly into the J input of FF_1 and through the inverter into the K input of FF_1. Thereby, the count sequence is started with FF_1 automatically preloaded (Q = 1). On the following clock pulse, the 1 bit is shifted into FF_2; since $\overline{Q}$ is logic-low, there is no output from the AND gate. The 1 bit is shifted next into FF_3, and there is still no output from the AND gate. However, when the flip-flops attain the 0001 state, the AND gate outputs a pulse, and the cycle previously described is then repeated.

Troubleshooting Ring-Counter Circuitry

Troubleshooting ring-counter circuitry, as in most types of digital networks, involves internal shorts, stuck highs, stuck lows, open circuits, open bonds, and external shorts, as depicted in Fig. 8-9. Open-circuit faults are associated with voltage activity, but no current activity. Short-circuit faults are associated with current activity, but no voltage activity. A dead driver is associated with no voltage activity and no current activity. A logic pulser is used to drive an input or any node. A logic probe is used to check voltage activity, and a current tracer is used to check current activity.

171

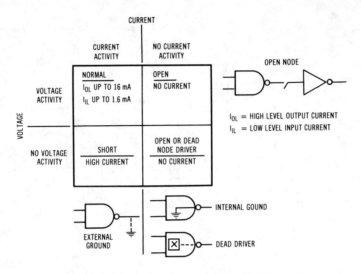

Fig. 8-9.   Voltage and current activity in digital circuitry.

SWITCH-TAIL (TWISTED-RING) COUNTER

A variation of the basic ring counter, called the *switch-tail counter*, *twisted-ring counter*, or *Johnson counter*, will also be encountered. In this configuration, the $\overline{Q}$ output of the last flip-flop is connected back to the J input of the first flip-flop (Fig. 8-10). This "switch" in output/input connections causes the counter to sequence through 2N different output states, where N is the number of flip-flops in the counter. A four-bit switch-tail counter has twice as many output states as a four-bit ring counter, although a straight 8421 binary counter has four times as many output states as a four-bit ring counter. Moreover, a switch-tail counter requires a decoder to output a separate signal for each of its 2N states.

Like a straight ring counter, a switch-tail counter can also be trapped into a self-perpetuating sequence of erroneous ("special") states; however, auxiliary logic can be utilized to force an erroneous sequence back to a normal sequence automatically. One method of implementing the correcting circuit is shown in Fig. 8-11A. A single AND gate is connected to the outputs of the last two flip-flops and to the K input of the first flip-flop. As a result, any of the eight possible "special" states will be forced back to the normal state within no more than five clock pulses. This method is used on switch-tail counters with an arbitrary number of flip-flops.

172

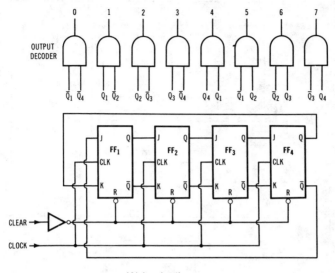

(A) Logic diagram.

(B) Truth table.

STATE	FLIP-FLOP			
	1	2	3	4
1	0	0	0	0
2	1	0	0	0
3	1	1	0	0
4	1	1	1	0
5	1	1	1	1
6	0	1	1	1
7	0	0	1	1
8	0	0	0	1
1	0	0	0	0

Fig. 8-10. A switch-tail (twisted-ring) counter.

Another method of implementing the correcting circuit is shown in Fig. 8-11B. The $\overline{Q}$ output from FF_2 is connected to the K input of FF_4, and the connection from $\overline{Q}$ of FF_3 to K of FF_4 is deleted. This error-correcting arrangement can be employed with switch-tail counters that have an arbitrary number of flip-flops. However, this method results in an output count of $2N-1$ states; in a four-bit switch-tail counter, the 0011 state will be followed by a 0000 state, and the 0001 state is eliminated.

TRUE BINARY COUNTERS

A *true binary counter* circuit differs from all forms of shift registers that have been described in that the flip-flops are in-

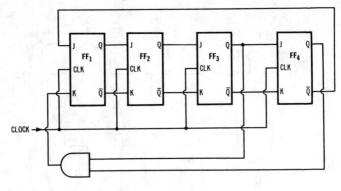

(A) One connection.

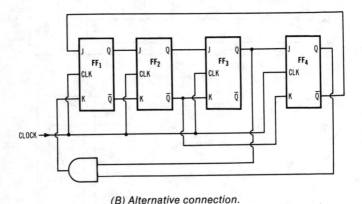

(B) Alternative connection.

Fig. 8-11. Mod-8 self-correcting switch-tail counter.

terconnected in a distinctive manner. Commercial counters are designed to output data in a specific pattern, or to maximize the number of output states that can be obtained with a given number of flip-flops. Most counters output data in terms of an 8421, 2'421, excess 3, or other commonly used binary code. The 8421 code is the familiar arrangement in which the binary digits (bits) have weights of 8, 4, 2, and 1 in order from the most significant bit (msb) to the least significant bit (lsb). In the 2'421 code, the weighting is 2, 4, 2, 1 from the msb position to the lsb position (Table 8-1). In the excess 3 code, 3 is added to a given decimal digit, and the result is converted to binary form (Table 8-2). The 2'421 and excess 3 codes are used when processing of the data requires obtaining the nines complement of the decimal digit that is represented by the code. (The nines comple-

Table 8-1. 2′421 Code

DECIMAL	2′421
0	0000
1	0001
2	0010
3	0011
4	0100
5	1011
6	1100
7	1101
8	1110
9	1111

Nines complement is obtained by taking ones complement.

Table 8-2. Excess 3 Code

DECIMAL	EXCESS 3 CODE
0	0011
1	0100
2	0101
3	0110
4	0111
5	1000
6	1001
7	1010
8	1011
9	1100

Nines complement is obtained by taking ones complement.

ment is the decimal digit that produces 9 when added to the given decimal digit.)

Many variations of the basic counter circuits are possible. All circuits employ JK, T, RS, or D-type flip-flops and can be classified into two fundamental categories, *asynchronous* counters (also known as *serial* or *ripple* counters) and *synchronous* counters (also known as *parallel* counters).

In a synchronous counter, all of the flip-flops change state simultaneously. In an asynchronous counter, one flip-flop changes state, and this change triggers a second flip-flop, which may then trigger a third flip-flop, and so on. Within these two fundamental categories, a counter can be configured to count to any desired binary number and then start the count sequence again. As mentioned previously, the modulus of a counter denotes the number of consecutive states through which the counter sequences before starting again. Thus, a modulo-n counter has n unique states. A binary counter may count up, count down, or both.

Asynchronous Counters

An asynchronous (binary ripple) counter is a fundamental configuration. For example, the arrangement of JK flip-flops in Fig. 8-12 will count *up* in binary 8421 code; the arrangement shown in Fig. 8-13 will count *down*. Trailing-edge-triggered flip-flops are employed. These counter arrangements are asynchronous because the output from the first flip-flop triggers the second flip-flop at its clock input; FF_3 is triggered from FF_2, and FF_3 in turn triggers FF_4. As noted earlier, a clock pulse applied at FF_1 will "ripple" from one flip-flop to the next; a counter in which this mode of propagation occurs is called a *ripple counter* or *series counter*.

With reference to the asynchronous up-counter configuration

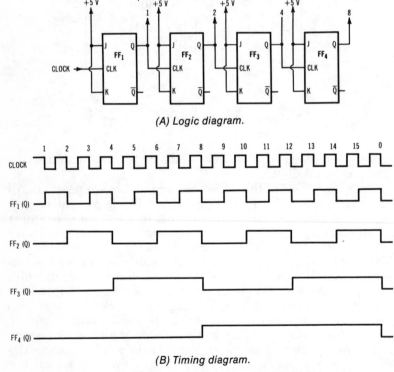

(A) Logic diagram.

(B) Timing diagram.

Fig. 8-12. Basic asynchronous up-counter.

(Fig. 8-12), observe that each JK flip-flop has its J and K inputs tied to a logic-high level. This causes the flip-flop to toggle (change state) each time a clock pulse is applied. Inasmuch as the output from one flip-flop is connected to the clock input of the next flip-flop, each successive flip-flop changes state half as often as the preceding flip-flop. That is, the clock input of a flip-flop must go from low to high and back to low again for its output to change state once. It is evident from the timing diagram in Fig. 8-12B that the divide-by-two action of each flip-flop produces logic-state changes according to the 8421 binary code. Note that the asynchronous down-counter (Fig. 8-13) employs the same operating principle, except that the $\overline{Q}$ output is utilized to cause the following flip-flop to toggle, and the code sequence is thereby reversed.

If the clock runs too fast, a counter will malfunction. Any counter has a maximum rate at which it can normally operate.

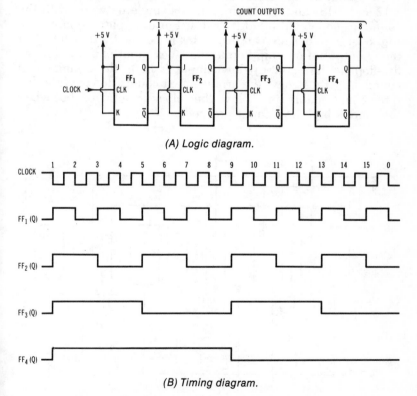

(A) Logic diagram.

(B) Timing diagram.

Fig. 8-13. Basic asynchronous down-counter.

177

For example, if each flip-flop in Fig. 8-13 has an inherent 25-nanosecond delay, the total delay from the occurrence of a trailing edge at the clock input of FF_1 to the completion of a state change in FF_4 will be 100 nanoseconds. Accordingly, the next clock-pulse trailing edge should not occur sooner than 100 ns. During the time between successive clock pulses, the flip-flops are changing states, and the output of the counter is incorrect. Although not all of the flip-flops must change states with each clock pulse, when the counter advances from 7 to 8 or when it recycles from 15 back to 0, all of the flip-flops necessarily change state; the clock signal must not recur before these state changes can be finalized. In this example, the maximum permissible clock frequency would correspond to a period of 100 ns; thus it would be 10 MHz.

Asynchronous Up/Down Counter

The count-up and count-down configurations shown in Figs. 8-12 and 8-13 can be combined into a single network, with the counting mode controlled by gates. For example, the circuit in Fig. 8-14 employs AND and OR gates to connect either the Q or the $\overline{Q}$ output of one flip-flop to the clock input of the following flip-flop. If the count-up/down input is high, the arrangement counts up; if it is low, the arrangement counts down. Note that if the count-enable line is low, the "clock is stopped," but when this line is high, the count proceeds.

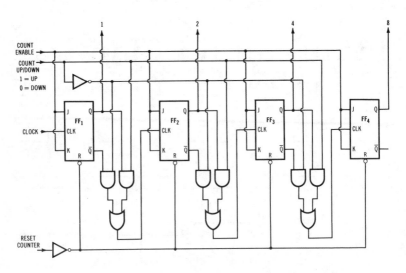

Fig. 8-14.   Example of asynchronous up/down counter.

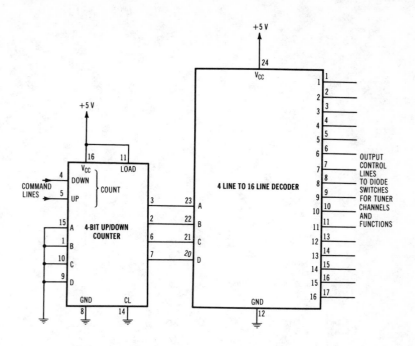

Fig. 8-15. Application of up/down counter in digital tv receiver.

Up/Down Counter in Digital TV

An example of an up/down counter and decoder application in a digital tv receiver is shown in Fig. 8-15. The decoder is an arrangement of gates that converts a binary coded input into a specific output. For example, if the counter outputs a 0001 signal, the decoder interprets this signal as a decimal 1 output. Next, if the counter outputs a 0010 signal, the decoder interprets this signal as a decimal 2 output. A 0011 signal is interpreted as a decimal 3 output, and so on. With four counter output lines, the decoder may provide up to 16 decimal output lines. In the example of Fig. 8-15, these decimal output lines correspond to tuner channels.

179

Chapter
9

Additional Counters and
Frequency Dividers

Additional types of counters, both synchronous and asynchronous, will be discussed in this chapter. Some practical applications of frequency dividers will be explained, and some troubleshooting pointers will be given.

SYNCHRONOUS COUNTERS

A synchronous counter is typified by the use of a common clock signal to trigger all of the flip-flops simultaneously. Usually, JK or type T flip-flops are used, as in asynchronous counters. The J and K inputs of each flip-flop are connected to the Q outputs of the preceding flip-flops through AND gates, as shown in Fig. 9-1. Any particular flip-flop will toggle when its associated AND gate brings its J and K inputs logic-high; this can occur only when all preceding flip-flops in the counter are logic-high. It is common practice to trigger synchronous counters by the leading edge of the clock pulse, instead of by the trailing edge.

An example of a 7-bit synchronous counter is shown in Fig. 9-1. It counts to 1111111 (127 in decimal notation). (Since it sequences through 128 states, including zero, this is a mod-128 counter.) It is instructive to consider the normal propagation delay for this configuration. If a JK flip-flop has a propagation delay of 25 ns and an AND gate has a propagation delay of 10 ns, the net delay is 35 ns, and a clock frequency of approximately 30 MHz can be utilized. Compare this frequency with the 10-MHz clock frequency used in an earlier example of a 4-bit asynchronous counter configured with the same flip-flops. In many applications, a synchronous counter provides another de-

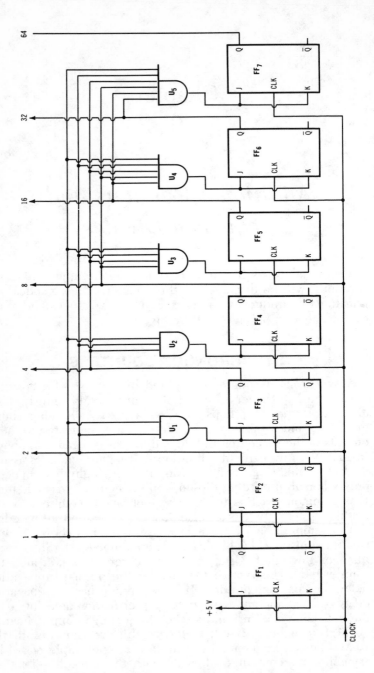

Fig. 9-1.   Example of 7-bit synchronous counter.

sirable feature in that all of the output lines change state simultaneously. That is, there are no intermediate states with incorrect counter outputs while the counter advances.

A synchronous counter does have certain limitations. The synchronous counter employs more logic gating and therefore is more complex and costly than its asynchronous counterpart. Observe also that AND gate U5 in Fig. 9-1 requires six inputs (seven inputs if an enable line is included). Thus, a counter of this type with a higher capacity would require a prohibitive number of inputs to any subsequent AND gate. Practical design tradeoffs influence the implementation of elaborate counter circuitry.

SYNCHRONOUS COUNTER
WITH RIPPLE CARRY

A simplified version of the straight synchronous counter, called a *ripple-carry counter,* is shown in Fig. 9-2. This configuration is synchronous in the sense that all flip-flops change state simultaneously. On the other hand, the connection between the J and K inputs of a particular flip-flop and the Q outputs of all preceding flip-flops is made through AND gates that are connected in series. Consequently, the propagation delay of the AND gates is cumulative, with the result that the maximum operating frequency is less than in the more elaborate version. Note, too, that for a higher count capacity, the propagation delay becomes proportionally greater, and this type of counter has a diminished speed advantage over an asynchronous version. Nevertheless, the ripple-carry counter often affords

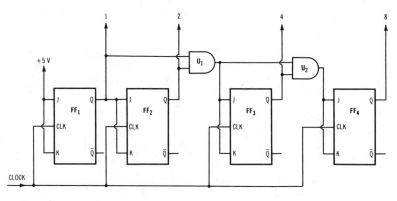

Fig. 9-2. Synchronous ripple-carry counter.

an attractive compromise between synchronous and asynchronous operation because its outputs change state simultaneously, and because its implementation is simpler than the straight synchronous counter.

BINARY CODED DECIMAL COUNTERS

A binary coded decimal (bcd) counter is a modulo-10 configuration; it counts to 10 (0 through 9), then automatically resets itself. This type of counter is widely used in calculator, computer, and similar arrangements where a decimal count is required. A typical configuration is shown in Fig. 9-3. Binary coded decimal counters may be either synchronous or asynchronous; special gating limits the count to 10 states in either

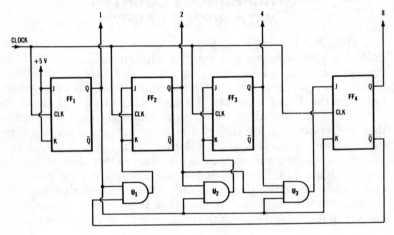

(A) Logic diagram.

INPUT BIT	OUTPUT
0	0000
1	0001
2	0010
3	0011
4	0100
5	0101
6	0110
7	0111
8	1000
9	1001
10	0000

(B) Count sequence.

Fig. 9-3. A binary coded decimal counter.

184

case. A bcd counter usually outputs 8421 code, although this practice is not universal.

The bcd count progression involves ten states, but the flip-flops have a potential capacity of 16 states, and some counters can be preset to any state from 0 to 15. Since a bcd counter will not count beyond 9, any preset value beyond 9 must be corrected and the counter returned to its normal sequence. For example, in one type of bcd counter, if the counter is preset to state 10, 11, 12, 13, 14, or 15, its internal circuitry will return to the normal count sequence within two clock cycles. This return can be shown by a *state diagram* (Fig. 9-4), which represents the progression of logic states for the device. For example, if the device happens to be preset to state 12, one clock pulse advances it to state 13, and the next clock pulse advances it to state 4, which is within the desired sequence.

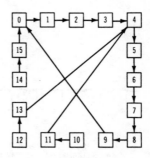

Fig. 9-4. State diagram for type 9310 bcd counter.

COUNTER IC PACKAGES

The four-bit binary counter IC in Fig. 9-5 has 16 output states. Asynchronous operation is provided, with a gated master reset line (MR_1 and MR_2 must both be high to drive all four Q outputs low). The flip-flops employ negative-edge triggering.

The decade counter IC in Fig. 9-6 is a synchronous arrangement that counts from 0 through 9 then automatically resets itself. Two dual reset inputs are provided; the R_0 reset input is used to reset the flip-flops to 0 asynchronously; the R_9 reset input is used to reset the flip-flops to 9 asynchronously. (In Fig. 9-6A, the gates that route the reset inputs to the proper flip-flops are considered to be internal to the flip-flops.) If frequency division by a factor of 10 and a symmetrical output are desired, output Q_D may be connected externally to input $\overline{CP}_A$. The input pulses are then applied to input $\overline{CP}_{BD}$, and a square wave at one-tenth of the input rate is obtained at output Q_A.

185

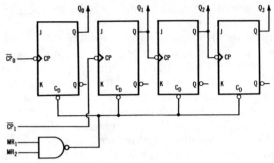

(A) Logic diagram.

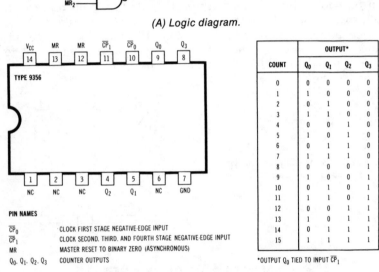

PIN NAMES

$\overline{CP}_0$	CLOCK FIRST STAGE NEGATIVE-EDGE INPUT
$\overline{CP}_1$	CLOCK SECOND, THIRD, AND FOURTH STAGE NEGATIVE-EDGE INPUT
MR	MASTER RESET TO BINARY ZERO (ASYNCHRONOUS)
Q_0, Q_1, Q_2, Q_3	COUNTER OUTPUTS

(B) Package pinout.

COUNT	OUTPUT*			
	Q_0	Q_1	Q_2	Q_3
0	0	0	0	0
1	1	0	0	0
2	0	1	0	0
3	1	1	0	0
4	0	0	1	0
5	1	0	1	0
6	0	1	1	0
7	1	1	1	0
8	0	0	0	1
9	1	0	0	1
10	0	1	0	1
11	1	1	0	1
12	0	0	1	1
13	1	0	1	1
14	0	1	1	1
15	1	1	1	1

*OUTPUT Q_0 TIED TO INPUT $\overline{CP}_1$

(C) Truth table.

Fig. 9-5.   A 4-bit binary counter IC.

EXAMPLES OF FREQUENCY DIVIDERS

An example of a divide-by-three frequency divider is shown in Fig. 9-7. Such a circuit might be used in a service-type color-bar generator. Let us assume that the Q output of FF_1 is logic-low and that the Q output of FF_2 is logic-low. The first input pulse will cause both flip-flops to change state because the set (S) and clear (C) inputs of FF_1 are logic-low, the S input of FF_2 is logic-high, and the C input of FF_2 is logic-low. (The effects of the levels at the S and C inputs are those for the flip-flops used in the particular device from which this example was taken.) The second input pulse causes FF_2 to change state be-

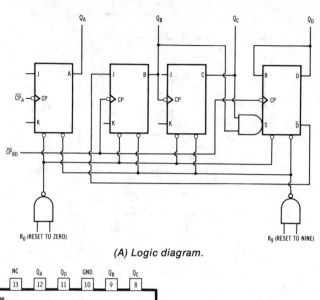

(A) Logic diagram.

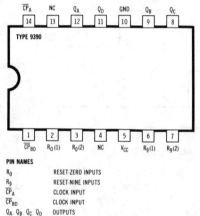

		OUTPUT*		
COUNT	Q_D	Q_C	Q_B	Q_A
0	0	0	0	0
1	0	0	0	1
2	0	0	1	0
3	0	0	1	1
4	0	1	0	0
5	0	1	0	1
6	0	1	1	0
7	0	1	1	1
8	1	0	0	0
9	1	0	0	1

*OUTPUT Q_A CONNECTED TO INPUT $\overline{CP}_{BD}$

PIN NAMES

R_0	RESET-ZERO INPUTS
R_9	RESET-NINE INPUTS
$\overline{CP}_A$	CLOCK INPUT
$\overline{CP}_{BD}$	CLOCK INPUT
Q_A, Q_B, Q_C, Q_D	OUTPUTS

(B) Package pinout.   *(C) Truth table.*

Fig. 9-6. A decade-counter IC.

cause C of FF_2 is high and S of FF_2 is low; (FF_1 is prevented from changing state by the high level at its S and C inputs). Now S and C of FF_1 are low, C of FF_2 is high, and S of FF_2 is low. The third input pulse causes FF_1 to change state, but FF_2 remains unchanged. These relationships are summarized in the timing diagram of Fig. 9-7B.

Consider next the divide-by-11 frequency divider shown in Fig. 9-8; a similar circuit might be used in a color-bar generator. Note carefully that this circuit is analyzed in terms of *negative*

187

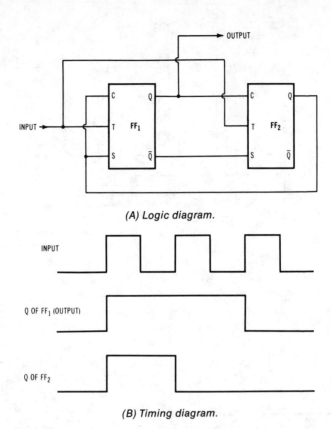

(A) Logic diagram.

(B) Timing diagram.

Fig. 9-7. A divide-by-three circuit.

logic. When a high (logic 0) level is applied to the preset (P) inputs of the flip-flops, the Q outputs go low (logic 1), and the $\overline{Q}$ outputs go high (logic 0). Gate G_1 produces a positive (logic 0) output only when all of its inputs are low (logic 1); this occurs only on the eleventh pulse count (state 10 of the flip-flops). The output pulse from G_1 is differentiated by C_1 and R_1, thereby producing a positive spike when the leading edge of the pulse occurs and a negative spike when the trailing edge of the pulse occurs. The negative spike cuts off Q_1 and produces a positive spike that is applied to the preset inputs of the flip-flops. The flip-flops are thus reset, and the 11-count sequence begins again.

TROUBLESHOOTING NOTES

Digital ICs fail about three-fourths of the time by open-circuiting at either the input or the output. (See Fig. 9-9.) Open

188

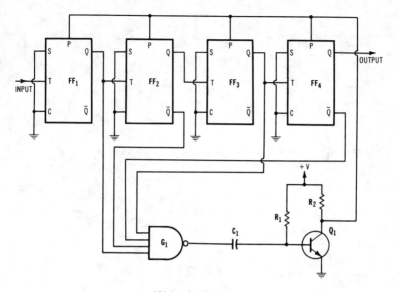

(A) Logic diagram.

NEGATIVE LOGIC
(HIGH = 0; LOW = 1)

(B) Truth table.

STATE	Q OUTPUTS			
	FF₄	FF₃	FF₂	FF₁
0	1	1	1	1
1	1	1	1	0
2	1	1	0	1
3	1	1	0	0
4	1	0	1	1
5	1	0	1	0
6	1	0	0	1
7	1	0	0	0
8	0	1	1	1
9	0	1	1	0
10	0	1	0	1

Fig. 9-8. A divide-by-eleven circuit.

circuits can be pinpointed by logic probes and pulsers. Short circuits and "stuck low" faults are generally analyzed to best advantage by current tracing (excessive current is usually present). However, if there is little or no current at a node, there is the probability of a dead driver (open output bond), or a lack of pulse activity in the circuit. In these situations, it is advisable to use a logic probe and pulser to close in on the trouble area, and then to follow up with a pulser and current tracer. Note that V_{cc} and ground lines tend to be very "noisy" with current spikes,

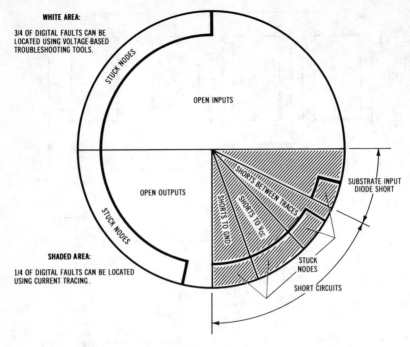

Fig. 9-9. Statistical summary of circuit failures. (*Courtesy Hewlett-Packard*)

although the voltage is virtually constant. Avoid tracing current on these lines, if possible. A logic pulser can be used to boost the current on the node under test and thereby facilitate analysis.

Counter troubleshooting is sometimes facilitated by the ability to inject a known number of pulses. With reference to Chart 9-1, a Hewlett-Packard 546A pulser can be "programmed" to output a desired number of pulses. There are six output modes, as shown in the left column of the chart. The pulses produced in programming the output are subtracted from the initial burst. The pulse button should be released *during* the final burst; the pulser will then complete the burst and turn itself off. The pulser includes an LED to indicate when it is operating. This also verifies the program mode and permits the counting of pulse bursts. For example, 432 output pulses can be obtained by programming four bursts of 100, three bursts of ten, and two single pulses. The operations required to accomplish this sequence with the Hewlett-Packard 546A pulser are shown in the right column of Chart 9-1.

190

Chart 9-1. Logic Pulser Output Programming Modes

OPERATION:		
PRESS AND RELEASE CODE BUTTON O		
PRESS AND LATCH CODE BUTTON Q		

OUTPUT MODES:		TO OUTPUT EXACTLY 432 PULSES:	
O	Single Pulse	1. 100 Hz Burst O Q	98
			100
Q	100 Hz Stream		100
			100
O Q	100 Hz Burst		400
		2. 10 Hz Burst O O O Q	6
O O Q	10 Hz Stream		10
			10
O O O Q	10 Hz Burst		430
		3. Single Pulse O	1
O O O O Q	1 Hz Stream	Single Pulse O	1
			432

DIGITAL CLOCK

A logic diagram for a digital clock is shown in Fig. 9-10. It is energized by a 1-Hz input from a time base (not shown) that divides the power-line frequency by 60. The arrangement in Fig. 9-10 provides either 12- or 24-hour readout, push-button advance of minutes and hours, and a switch to stop the seconds display when the clock is set against a standard such as WWV.

In the following discussion, reference is made to Fig. 9-10 for the overall operation of the clock, and to Fig. 9-11 for details of the individual ICs. With reference to Fig. 9-10, IC_{13} receives the 1-Hz trigger from the external time base; IC_{13} (Fig. 9-11A) is gated on and off by commands from the "seconds stop" switch. This feature permits precise setting of the time display. From IC_{13}, the 1-Hz pulse is applied to IC_7 (Fig. 9-11B), which is a divide-by-10 circuit with binary coded decimal (bcd) output. This output is applied to IC_1, which converts the bcd signals into the code for a seven-segment display. Integrated circuit IC_1 contains, in addition to its logic circuitry, seven transistors that, when turned on, connect the LED display segments to ground (through protective resistors). After the numeral 9 is displayed, pin 11 of IC_7 (part of the bcd output) applies to IC_8 (Fig. 9-11C) a negative-going pulse that advances IC_8 by one digit.

Integrated circuit IC_8 is a divide-by-six circuit, whose bcd output is decoded by IC_2 to display the numerals 0 through 5. After numeral 5 is displayed, a negative-going pulse is supplied from pin 9 of IC_8 to IC_9 (Fig. 9-11B) for the minutes display, and

191

so on. Integrated circuits IC_7, IC_9, and IC_{11} are all divide-by-10 devices; IC_8, IC_{10}, and IC_{12} are divide-by-six devices.

The bcd outputs of IC_{11} and IC_{12} must generate numerals which do not reset to zero after 10 but continue on to 11 and 12, and then reset to 01 rather than 00. This mode of operation is accomplished as follows. Device IC_{13} (Fig. 9-11A) is wired in part to form a three-input AND gate (all three inputs must be logic-high to give a logic-high output). The output from this gate resets IC_{11} and IC_{12} back to numeral 0. This will occur only when the A output of IC_{12} (Fig. 9-11C) is logic-high, corresponding to the numeral 1 in the displayed number "12," and when the B output of IC_{11} is logic-high, corresponding to the numeral 2 in the displayed number "12," and when the A output of IC_{11} just goes logic-high (in an attempt to display the number "13"). At this precise moment, the AND circuit activates the resets, returning IC_{11} and IC_{12} to zero, which in turn shuts off the inputs to the AND circuit and shuts off the reset pulse before the A output of IC_{11} completely reaches a logic-high level, thereby not interfering with this function. This permits the A output to complete its change of state to logic-high to display the number "01." (The capacitor coupling between IC_{10} and IC_{11} serves to hold a charge during the foregoing operations.)

Push-button advance of the minutes and hours display is accomplished by "processing" the button inputs with a pair of RS flip-flops, using IC_{14} to remove the bounce and send the resultant pulses through capacitors to the inputs of IC_9 and IC_{11}. A debouncing circuit serves to lock in on a leading pulse, and to lock out any closely lagging pulses. Thereby, the sometimes uncertain contact waveform provided by a mechanical switch is "cleaned up" to eliminate possible spurious responses in the digital system.

Preliminary analysis of a digital system is often facilitated by the use of block diagrams, such as those in Fig. 9-12. The first block diagram (Fig. 9-12A) is generalized and provides a simplified summary of system operation with an elementary flow diagram. The second block diagram (Fig. 9-12B) includes the principal devices that are employed in the various sections. For example, it shows that the reference 60-Hz frequency is applied to a ripple counter that contains six flip-flops. Output from this divide-by-60 counter is applied to a 10×6 shift counter with eight flip-flops. (A shift counter is a shift register in which the first stage, through logic feedback, produces a pattern of ones or zeros [ring code] as a function of the state of the other stages in the register. A 10×6 shift counter employs a mod-10 counter

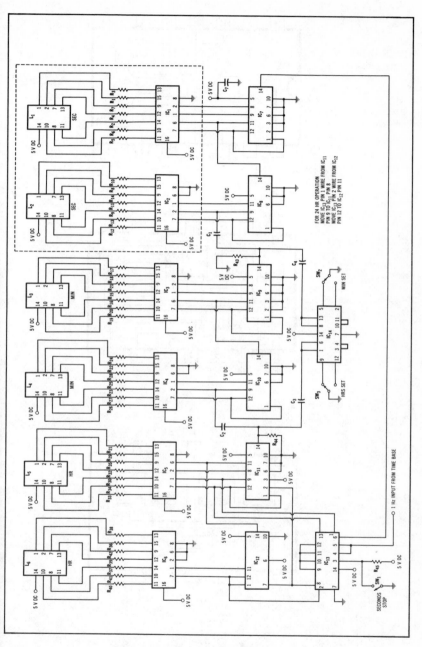

Fig. 9-10.   Diagram of a digital clock.

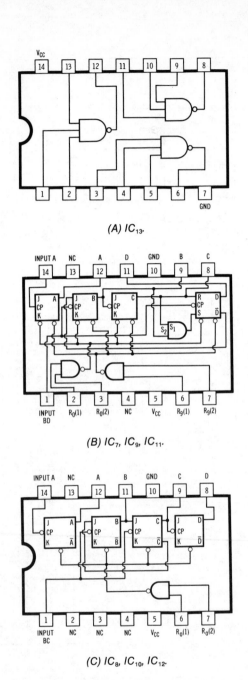

(A) IC_{13}.

(B) IC_7, IC_9, IC_{11}.

(C) IC_8, IC_{10}, IC_{12}.

Fig. 9-11. Logic diagrams of ICs in Fig. 9-10.

194

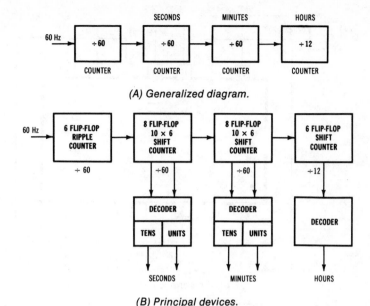

(A) Generalized diagram.

(B) Principal devices.

Fig. 9-12. Block diagrams of digital clock.

and a mod-6 counter.) A third block diagram, with additional circuit details, may be developed if desired.

COUNTERS IN HOME APPLIANCES

A digital counter is often the nucleus of a logic-operated home appliance. The counter may be required to "go back to zero" at some arbitrary point, and the technician needs to know how standard decade counters are programmed in this manner. For example, the widely used 7490 decade counter comprises divide-by-2 and divide-by-5 sections in the same IC package. If the pins are connected as shown in Fig. 9-13, the counter operates as a divide-by-10 device. This counter is triggered by the trailing edge of the applied pulse. In normal operation, pins 11, 8, 9, and 12 output bcd counts. The counter automatically resets to 0 following the 9 count.

Suppose that an appliance application requires resetting at the second count, so that the output is 0, 1, 0, 1, 0, and so on. In this case, pins 2 and 3 are disconnected from ground; pin 2 is connected to pin 9, and pin 3 is left "floating." Again, suppose that the appliance application requires resetting at the third count, so that the output is 0, 1, 2, 0, 1, 2, and so on. This mode

195

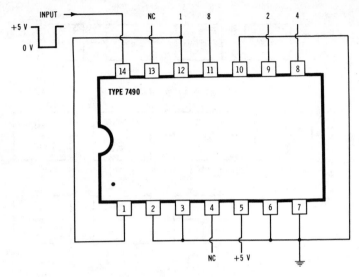

(A) Pin connection.

COUNT	OUTPUTS			
	11	8	9	12
0	0	0	0	0
1	0	0	0	1
2	0	0	1	0
3	0	0	1	1
4	0	1	0	0
5	0	1	0	1
6	0	1	1	0
7	0	1	1	1
8	1	0	0	0
9	1	0	0	1

(B) Output states.

Fig. 9-13.   Division by 10 with 7490 IC.

of operation is obtained by connecting pin 2 to pin 9, as before, and connecting pin 3 to pin 12. To obtain resetting on the fourth count, so that the output is 0, 1, 2, 3, 0, 1, 2, 3, and so on, pin 2 or pin 3 is connected to pin 8, with the other pin left "floating." (Pins 2 and 3 are called the reset pins.) To obtain resetting on the fifth count, one reset pin is connected to pin 8, and the other reset pin is connected to pin 12. To obtain resetting on the sixth count, one reset pin is connected to pin 8, and the other reset pin is connected to pin 9.

When an application requires resetting after the seventh count, a supplementary 3-input AND gate is employed. Pins 8, 9, and 12 are connected to the inputs of the AND gate; the output

196

from the AND gate is connected to pin 2, with pin 3 left "floating." When the application features more than one decimal digit, AND gates are utilized to respond to the states of several counter output lines. As an illustration, an application may require counting from 00 to 35 and then resetting to 00 at the 36th count. This mode of operation is accomplished as shown in Fig. 9-14. Division by 77 (resetting at the 77th count) may be obtained as depicted in Fig. 9-15.

In a more elaborate application, division by a number such as 777 might be required; the count would proceed from 000 to 776 and then back to 000. This mode of operation is obtained by means of the interconnections seen in Fig. 9-16.

Sometimes the counters can be programmed for two-digit counts without supplementary AND gates. For example, consider the operation of 7490 counters in a digital clock circuit. Here, the next count after 59 is not 60, but return to 00. This mode of operation is obtained as depicted in Fig. 9-17. This is a divide-by-60 counter in which the units counter automatically resets to 0 after count 9, whereas the tens counter resets on the 6 count.

FREQUENCY DIVISION IN A
DIGITAL COLOR-BAR GENERATOR

A diagram of a digital color-bar generator is shown in Fig. 9-18. This generator provides keyed-rainbow, crosshatch, vertical-line, horizontal-line, and white-dot outputs.

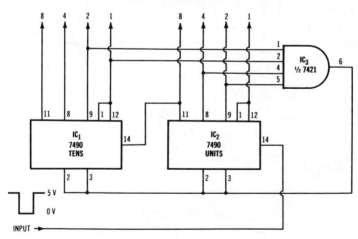

Fig. 9-14. Method of dividing by 36.

197

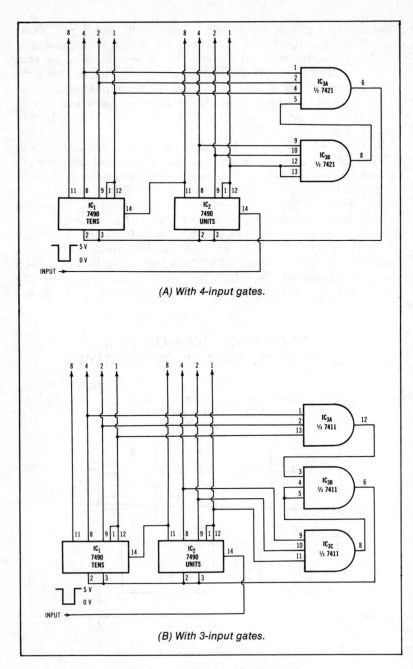

(A) With 4-input gates.

(B) With 3-input gates.

Fig. 9-15. Methods of dividing by 77.

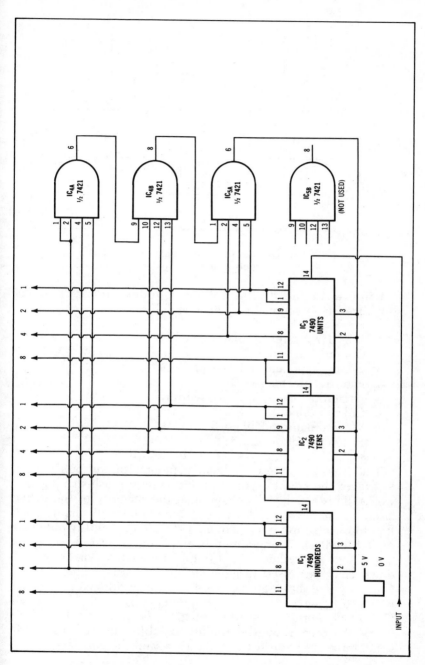

Fig. 9-16. Three counters connected to divide by 777.

199

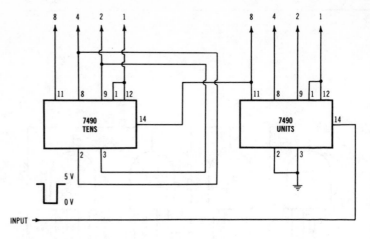

Fig. 9-17. Two counters interconnected to divide by 60 without supplementary AND gates.

Circuits in the digital color-bar generator may be sectionalized in functional sequence as follows:

1. 189-kHz master oscillator
2. Countdown chain
3. Gating logic and pulse-shaping networks
4. Pattern selector switch
5. Digital-to-analog mixer/converter
6. Color oscillator
7. Vhf oscillator/modulator
8. Power supply

The 189-kHz master oscillator generates timing pulses used to trigger the countdown chain and certain pulse-shaping networks. Thirteen flip-flops comprise the majority of the active circuitry in the countdown (frequency-division) chain. Each flip-flop has an inherent frequency-division capability of two. A digital feedback configuration is employed to scale by integers other than 2. Predetermined output frequencies of the countdown chain are used to generate the sync and video signals. Correct pulse shaping is accomplished by additional processing with gating logic and pulse-shaping networks. Gates G_1 through G_4 and G_6 perform the AND function; G_5 is an OR gate. The pulse-shaping networks are monostable multivibrators for generating the vertical lines, the dots, and the color bars.

200

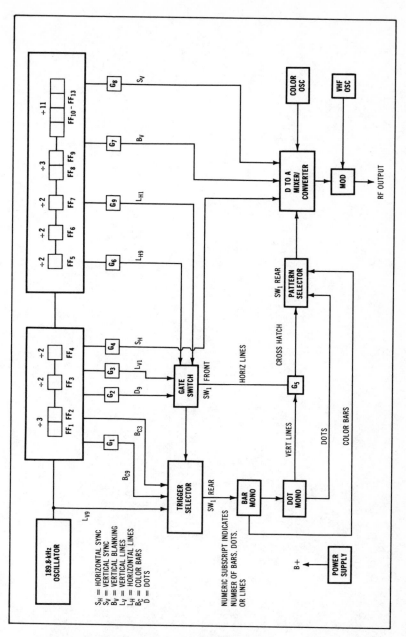

Fig. 9-18. Simplified block diagram of a service-type color-bar generator. (*Courtesy B&K Precision*)

Chapter
10

Encoders and Decoders

As noted previously, a variety of codes is employed in the operation of digital logic networks. An encoder is a device that is capable of translating from one method of expression to another method of expression. The arrangement in Fig. 10-1 is for changing decimal keystrokes into a binary coded decimal (bcd) output. For example, in bcd code, the decimal digit 7 corresponds to 0111. Therefore, the "7" keystroke applies logic-high levels to the 4, 2, and 1 OR gates. Note that only one decimal digit may be applied at a given time to the encoder input. This type of encoder is generally built into a microcomputer keyboard.

PRIORITY ENCODER

Another widely used encoder configuration is called a *priority encoder*. It is often utilized to control the access of peripheral devices (such as keyboards and printers) to a computer input/output channel so that a device that has the highest assigned priority gains access to the channel before any device with a lower assigned priority. A typical priority encoder may have eight input lines and three output lines, as shown in Fig. 10-2. Whenever one input line, such as $\overline{6}$ (6 active low), is enabled by the peripheral device connected to it, the priority encoder outputs a binary 6 (110) to the computer. When two or more input lines, such as $\overline{3}$, $\overline{6}$, and $\overline{7}$, are enabled, the priority encoder outputs a binary count corresponding to the highest-order line—in this example, $\overline{7}$. If only one input line is activated at a time, the priority encoder responds as an 8-line–to–binary encoder.

A commercial priority encoder IC package is shown in Fig.

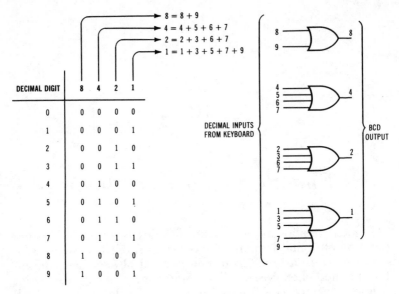

Fig. 10-1. A decimal-to-bcd encoder.

10-3. The logic diagram for this device is similar to Fig. 10-2, except that the outputs are all active low. When the enable input ($\overline{EI}$) is logic-high, all outputs go high. When input $\overline{EI}$ is logic-low, the eight input signals are encoded as listed in Fig. 10-3B. Observe that this configuration provides an enable output ($\overline{EO}$), a group-select ($\overline{GS}$) output, and three address outputs ($\overline{A}_0$, $\overline{A}_1$, and $\overline{A}_2$). Note also that the address outputs produce binary number equivalents of the decimally numbered inputs. For example, if the $\overline{5}$ input is pulsed low, the address outputs produce a signal that corresponds to $A_0A_1A_2 = 101$.

BASIC DECODERS

A decoder is a device for translating a combination of signals into one signal that represents the combination. As an illustration, a binary-to-decimal decoder is depicted in Fig. 10-4. This arrangement decodes the numbers 0 through 9, and also the numbers 30 and 31. Thus, an input of 00110 produces an output from AND gate U_7; an input of 11111 produces an output from AND gate U_{12}. A total of 32 AND gates would be required to decode all numbers possible from a 5-bit code; in other words, 2^5 AND gates would be required. In the example of Fig. 10-4, the decoder is activated from a shift register that has five flip-flops. One AND gate is utilized for each number that is to be decoded.

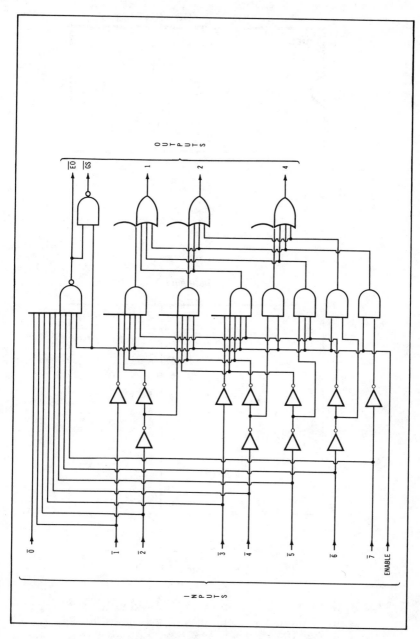

Fig. 10-2.   Logic diagram of a priority encoder. (*Courtesy Hewlett-Packard*)

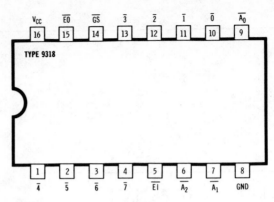

PIN NAMES

$\overline{0}$ TO $\overline{7}$	PRIORITY INPUTS (ACTIVE LOW)
$\overline{EI}$	ENABLE INPUT (ACTIVE LOW)
$\overline{EO}$	ENABLE OUTPUT (ACTIVE LOW)
$\overline{GS}$	GROUP SELECT OUTPUT (ACTIVE LOW)
$\overline{A_0}$, $\overline{A_1}$, $\overline{A_2}$	ADDRESS OUTPUTS (ACTIVE LOW)

(A) Package pinout.

$\overline{EI}$	$\overline{0}$	$\overline{1}$	$\overline{2}$	$\overline{3}$	$\overline{4}$	$\overline{5}$	$\overline{6}$	$\overline{7}$	$\overline{GS}$	$\overline{A_0}$	$\overline{A_1}$	$\overline{A_2}$	$\overline{EO}$
1	X	X	X	X	X	X	X	X	1	1	1	1	1
0	1	1	1	1	1	1	1	1	1	1	1	1	0
0	X	X	X	X	X	X	X	0	0	0	0	0	1
0	X	X	X	X	X	X	0	1	0	1	0	0	1
0	X	X	X	X	X	0	1	1	0	0	1	0	1
0	X	X	X	X	0	1	1	1	0	1	1	0	1
0	X	X	X	0	1	1	1	1	0	0	0	1	1
0	X	X	0	1	1	1	1	1	0	1	0	1	1
0	X	0	1	1	1	1	1	1	0	0	1	1	1
0	0	1	1	1	1	1	1	1	0	1	1	1	1

X = "DON'T CARE"

(B) Truth table.

Fig. 10-3. Eight-input priority encoder.

That is, when a particular number is stored in the shift register, the output of the corresponding AND gate goes to logic 1.

Note in Fig. 10-4 that for each binary number that is stored, every flip-flop has either its Q output or its $\overline{Q}$ output logic-high.

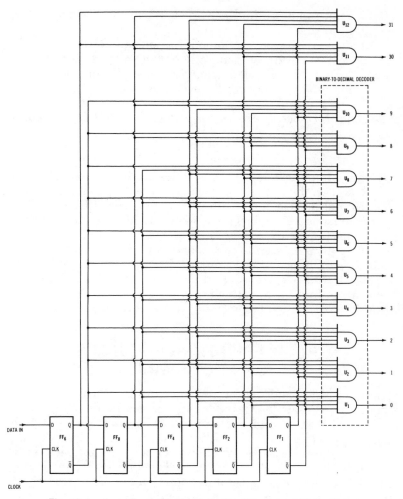

Fig. 10-4. A configuration for decoding binary-coded numbers.
(*Courtesy Hewlett-Packard*)

For the decimal number 31 (binary 11111), the Q output of each
flip-flop is logic-high. For the decimal number 0 (binary 00000),
the $\overline{Q}$ output of each flip-flop is logic-high. Thus, to decode the
number 31, the five inputs to AND gate U_{12} must be the five Q
outputs of the shift register. To decode the number 0, the five $\overline{Q}$
outputs provide the inputs to AND gate U_1. If the decimal
number 17 (binary 10001) is to be decoded, the Q outputs from
the first and last flip-flops and the $\overline{Q}$ outputs from the middle
three flip-flops are inputted to an AND gate (not shown in Fig.

10-4). This same method can be used to decode any other number.

Commercial designs of decoders translate binary, binary-coded decimal (bcd), or some other standard code into an uncoded form. The following example shows the distinction between the binary code and the bcd code:

Decimal Number	Binary Equivalent	BCD Equivalent
		(3) (1)
31	11111	0011 0001

Common applications of these codes are in the conversion of bcd data from the arithmetic circuits of a digital calculator into a decimal output that serves to turn on one of ten LEDs in a numerical display. Decoders are also commonly used to convert a group of four binary address bits into an output signal on any one of 16 lines in order to access one of 16 digital words that are stored in a random-access memory (RAM). One particular category of IC decoders is termed *decoder/drivers*; these devices have an output stage following the decoder so that a lamp, solid-state display, relay, or other device can be driven.

TROUBLESHOOTING ENCODERS
AND DECODERS

Logic pulsers and logic probes are the most useful troubleshooting instruments in encoder and decoder circuitry. As previously discussed, digital circuitry may be implemented with TTL or CMOS devices. Accordingly, it is often necessary to use a logic probe that can be switched to either TTL or CMOS indication, as illustrated in Fig. 10-5. The logic-high, logic-low, bad-level, and pulse indications shown in the diagram apply to both TTL and CMOS circuitry. Similarly, it is often necessary to use a logic pulser that can be switched to either TTL or CMOS operation.

Observe the logic-pulser output waveforms depicted in Fig. 10-6. The pulser provides a high-current narrow pulse for TTL tests, and a low-current wide pulse for CMOS tests. The pulser automatically drives a logic-low terminal high and drives a logic-high terminal low. The pulse widths are maintained within limits to ensure that devices under test cannot be damaged by the application of excessive energy. As noted previously, the pulser can be programmed to provide a chosen number of pulses, pulse bursts, or pulse trains.

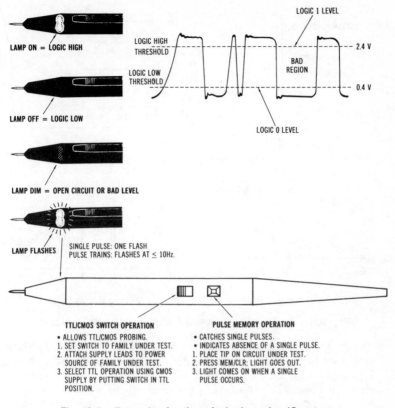

LAMP ON = LOGIC HIGH

LAMP OFF = LOGIC LOW

LAMP DIM = OPEN CIRCUIT OR BAD LEVEL

LAMP FLASHES | SINGLE PULSE: ONE FLASH
PULSE TRAINS: FLASHES AT ≤ 10Hz.

LOGIC HIGH THRESHOLD
LOGIC LOW THRESHOLD
LOGIC 1 LEVEL
BAD REGION — 2.4 V
— 0.4 V
LOGIC 0 LEVEL

TTL/CMOS SWITCH OPERATION
• ALLOWS TTL/CMOS PROBING.
1. SET SWITCH TO FAMILY UNDER TEST.
2. ATTACH SUPPLY LEADS TO POWER SOURCE OF FAMILY UNDER TEST.
3. SELECT TTL OPERATION USING CMOS SUPPLY BY PUTTING SWITCH IN TTL POSITION.

PULSE MEMORY OPERATION
• CATCHES SINGLE PULSES.
• INDICATES ABSENCE OF A SINGLE PULSE.
1. PLACE TIP ON CIRCUIT UNDER TEST.
2. PRESS MEM/CLR; LIGHT GOES OUT.
3. LIGHT COMES ON WHEN A SINGLE PULSE OCCURS.

Fig. 10-5.   Example of action of a logic probe. (*Courtesy Hewlett-Packard*)

Consider next the trouble symptom produced by an open bond in an AND gate output (Fig. 10-7). When test pulses are applied to the inputs of U_1, the output terminal (B) is "stuck at" a bad level. (This bad level will be interpreted as a logic-high state by U_3 if TTL or DTL circuitry is employed.) In this situation, a check of U_3 shows normal output when its inputs are pulsed. Therefore, the conclusion is that U_1 is defective and should be replaced.

2-TO-4–LINE DECODER

A simple 2-line–to–4-line IC decoder circuit is shown in Fig. 10-8. Inasmuch as the two inputs (A and B) together can have only the four different states shown in the truth table (binary code 00 to 11), a 2-input decoder can have only four unique

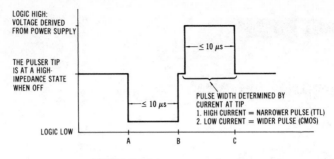

(A) Pulsing into an open circuit.

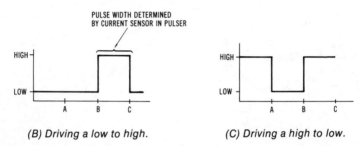

(B) Driving a low to high.   *(C) Driving a high to low.*

Fig. 10-6. Logic-pulser waveforms. (*Courtesy Hewlett-Packard*)

outputs. A different output gate is enabled by each binary state. There can be as many output gates as there are input states: a 3-input decoder can have eight unique outputs; a 4-input decoder can have 16 unique outputs; and so on.

STROBE LINES

A *strobe line* is used in various decoder arrangements. When the strobe line is in one logic state, all of the output lines are disabled; when it is in the other logic state, data output can proceed. A strobe pulse is typically a narrow pulse which is automatically generated when required. Fig. 10-9 shows an example of a bcd-to-decimal decoder with a strobe input. As explained above, the four inputs could be decoded into 16 unique outputs; however, the bcd code employs only the 10 binary counts from 0000 to 1001, and, therefore, this decoder has only 10 output gates. Accordingly, there is a possibility of false data processing in the elementary arrangement; that is, if an illegal code happens to be applied, the decoder would respond with a false output. (For example, input code 1010 would cause output lines 2 and 8 to be logic high.) Hence, five-input AND

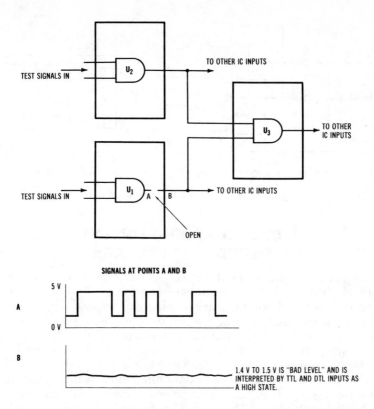

Fig. 10-7. Trouble symptom produced by an open bond in a gate output.
(*Courtesy Hewlett-Packard*)

gates are utilized, with a strobe input to each; unless the strobe line goes high, no output can occur. The strobe line by itself does not prevent inputting of false data, but it is used to enable the decoder only when the input data are known to be valid.

EXAMPLE OF 1-OUT-OF-10 DECODER/DRIVER IC PACKAGE

An example of a 1-out-of-10 decoder driver is shown in Fig. 10-10. The device accepts bcd inputs and provides output states to drive 10-digit incandescent readout displays. All outputs normally remain off for all invalid conditions, as shown in the truth table (Fig. 10-10C). For example, if an illegal code such as 1010 happens to be inputted, all of the active-low outputs go logic-high (see also Fig. 10-10D).

211

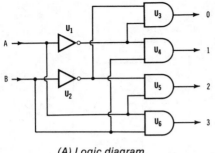

INPUTS		LOGIC 1 OUTPUT
A	B	
0	0	0
0	1	1
1	0	2
1	1	3

(A) Logic diagram.　　　　　(B) Truth table.

Fig. 10-8. Example of 2-line–to–4-line decoder.

BCD TO SEVEN-SEGMENT DECODER IC PACKAGE

A bcd to seven-segment decoder IC package is shown in Fig. 10-11. It is used to drive a seven-segment readout device (Fig. 10-12). Decimal numerals are displayed by energizing from two to seven segments in the readout device. Thus, seven decoder outputs are provided, and these outputs are selected by four coded inputs. For example, a 1 readout is obtained by applying a 0001 input to the decoder; a 9 readout is obtained by applying a 1001 input. Five additional symbolic readouts may be used in particular applications. The seven-segment format may be utilized in LED, LCD, gas-discharge, and filamentary readout devices.

In addition to the numerical inputs, the device in Fig. 10-11 has blanking and test inputs. If the $\overline{\text{RBI}}$ input is low and the bcd inputs are low (state 0000), outputs a through g are held low (the display is blanked); if $\overline{\text{RBI}}$ is high and the input is 0000, outputs a through f go high so that a zero is displayed. Terminal 4 serves the dual function of blanking input and ripple-blanking output; hence, it is labeled $\overline{\text{BI/RBO}}$. If this terminal is driven low, outputs a through g go low, and the display is blanked. If $\overline{\text{RBI}}$ is low and the input is 0000, the display is blanked, and $\overline{\text{BI/RBO}}$ goes low to pass the blanking signal on to another device. The lamp test ($\overline{\text{LT}}$) input activates outputs a through g when it is driven low while $\overline{\text{BI/RBO}}$ is high.

DECODER WITH ADDRESS INPUTS

Another decoder IC package is shown in Fig. 10-13. This is a 1-out-of-16 configuration employing four inputs and 16 outputs

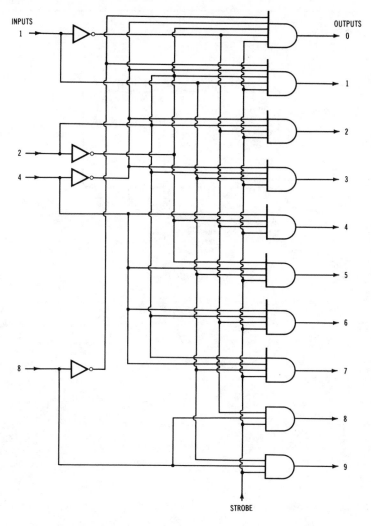

Fig. 10-9.   Example of a bcd to decimal decoder. (*Courtesy Hewlett-Packard*)

(the maximum possible number). The inputs are designated as *address* inputs; an address is a binary number that corresponds to a location where digital information is stored in a memory. An enable function is provided, similar to the strobe line previously described. This enable line is ANDed to a pair of inputs for application flexibility.

213

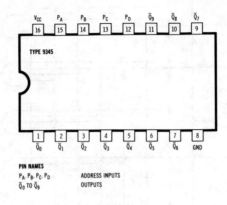

(A) Package pinout.

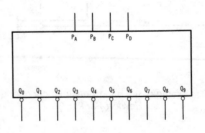

(B) Logic symbol.

Fig. 10-10. A 1-out-of-10

INPUTS				OUTPUTS									
P_D	P_C	P_B	P_A	$\bar{Q}_0$	$\bar{Q}_1$	$\bar{Q}_2$	$\bar{Q}_3$	$\bar{Q}_4$	$\bar{Q}_5$	$\bar{Q}_6$	$\bar{Q}_7$	$\bar{Q}_8$	$\bar{Q}_9$
0	0	0	0	0	1	1	1	1	1	1	1	1	1
0	0	0	1	1	0	1	1	1	1	1	1	1	1
0	0	1	0	1	1	0	1	1	1	1	1	1	1
0	0	1	1	1	1	1	0	1	1	1	1	1	1
0	1	0	0	1	1	1	1	0	1	1	1	1	1
0	1	0	1	1	1	1	1	1	0	1	1	1	1
0	1	1	0	1	1	1	1	1	1	0	1	1	1
0	1	1	1	1	1	1	1	1	1	1	0	1	1
1	0	0	0	1	1	1	1	1	1	1	1	0	1
1	0	0	1	1	1	1	1	1	1	1	1	1	0
1	0	1	0	1	1	1	1	1	1	1	1	1	1
1	0	1	1	1	1	1	1	1	1	1	1	1	1
1	1	0	0	1	1	1	1	1	1	1	1	1	1
1	1	0	1	1	1	1	1	1	1	1	1	1	1
1	1	1	0	1	1	1	1	1	1	1	1	1	1
1	1	1	1	1	1	1	1	1	1	1	1	1	1

(C) Truth table.

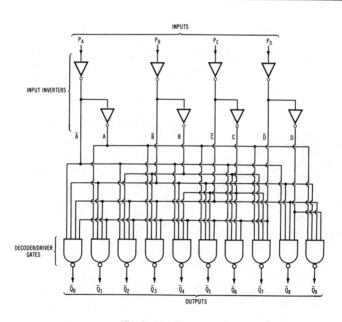

(D) Logic diagram.

decoder/driver IC package.

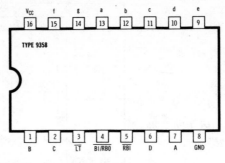

PIN NAMES

A, B, C, D	BCD INPUTS
RBI	RIPPLE-BLANKING INPUT
BI/RBO	BLANKING INPUT/RIPPLE- BLANKING OUTPUT
a, b, c, d, e, f, g	SEVEN-SEGMENT OUTPUTS
LT	LAMP TEST

(A) Package pinout.

NUMERAL OR FUNCTION	INPUTS						BI/RBO	OUTPUTS						
	LT	RBI	D	C	B	A		a	b	c	d	e	f	g
0	1	1	0	0	0	0	1	1	1	1	1	1	1	0
1	1	X	0	0	0	1	1	0	1	1	0	0	0	0
2	1	X	0	0	1	0	1	1	1	0	1	1	0	1
3	1	X	0	0	1	1	1	1	1	1	1	0	0	1
4	1	X	0	1	0	0	1	0	1	1	0	0	1	1
5	1	X	0	1	0	1	1	1	0	1	1	0	1	1
6	1	X	0	1	1	0	1	0	0	1	1	1	1	1
7	1	X	0	1	1	1	1	1	1	1	0	0	0	0
8	1	X	1	0	0	0	1	1	1	1	1	1	1	1
9	1	X	1	0	0	1	1	1	1	1	0	0	1	1
10	1	X	1	0	1	0	1	0	0	0	1	1	0	1
11	1	X	1	0	1	1	1	0	0	1	1	0	0	1
12	1	X	1	1	0	0	1	0	1	0	0	0	1	1
13	1	X	1	1	0	1	1	1	0	0	1	0	1	1
14	1	X	1	1	1	0	1	0	0	0	1	1	1	1
15	1	X	1	1	1	1	1	0	0	0	0	0	0	0
BI	X	X	X	X	X	X	0	0	0	0	0	0	0	0
RBI	1	0	0	0	0	0	0	0	0	0	0	0	0	0
LT	0	X	X	X	X	X	1	1	1	1	1	1	1	1

X = "DON'T CARE"

(B) Truth table.

Fig. 10-11.   A bcd to seven-segment decoder.

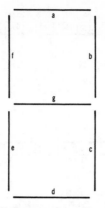

(A) Segment identification.

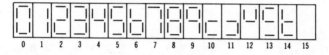

(B) Displayed symbols.

Fig. 10-12. Seven-segment readout.

DECIMAL-TO-BINARY ENCODER

Decimal-to-binary encoding (code conversion) is often accomplished by means of a diode matrix, a form of read-only memory, as shown in Fig. 10-14. The read-only memory (ROM) inputs decimal signals and outputs binary signals. It comprises 10 input lines and four output lines, interconnected by means of diodes. Each diode operates as a "one-way" device between an input line and an output line to channel an incoming digital pulse into a single output line, and to block "sneak currents" that otherwise would be able to pass from the output line back into the input lines.

The encoder shown in Fig. 10-14 operates as an interface between a keyboard (keypad) and the four output lines. Observe that when the "6" key is closed, current will pass from the battery through two diodes and into the B and C output lines. In turn, the binary number 0110 is outputted. Only one key should be depressed at a time; if two or more keys are closed simultaneously, additional output lines will be activated, and the output state will be meaningless.

217

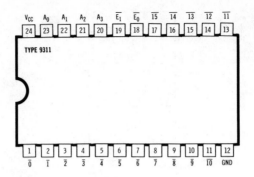

PIN NAMES

A_0, A_1, A_2, A_3	ADDRESS INPUTS
$\overline{E}_0, \overline{E}_1$	AND ENABLE INPUTS (ACTIVE LOW)
$\overline{0}$ TO $\overline{15}$	OUTPUTS (ACTIVE LOW)

(A) Package pinout.

$\overline{E}_0$	$\overline{E}_1$	A_0	A_1	A_2	A_3	$\overline{0}$	$\overline{1}$	$\overline{2}$	$\overline{3}$	$\overline{4}$	$\overline{5}$	$\overline{6}$	$\overline{7}$	$\overline{8}$	$\overline{9}$	$\overline{10}$	$\overline{11}$	$\overline{12}$	$\overline{13}$	$\overline{14}$	$\overline{15}$
1	1	X	X	X	X	1	1	1	1	1	1	1	1	1	1	1	1	1	1	1	1
1	0	X	X	X	X	1	1	1	1	1	1	1	1	1	1	1	1	1	1	1	1
0	1	X	X	X	X	1	1	1	1	1	1	1	1	1	1	1	1	1	1	1	1
0	0	0	0	0	0	0	1	1	1	1	1	1	1	1	1	1	1	1	1	1	1
0	0	1	0	0	0	1	0	1	1	1	1	1	1	1	1	1	1	1	1	1	1
0	0	0	1	0	0	1	1	0	1	1	1	1	1	1	1	1	1	1	1	1	1
0	0	1	1	0	0	1	1	1	0	1	1	1	1	1	1	1	1	1	1	1	1
0	0	0	0	1	0	1	1	1	1	0	1	1	1	1	1	1	1	1	1	1	1
0	0	1	0	1	0	1	1	1	1	1	0	1	1	1	1	1	1	1	1	1	1
0	0	0	1	1	0	1	1	1	1	1	1	0	1	1	1	1	1	1	1	1	1
0	0	1	1	1	0	1	1	1	1	1	1	1	0	1	1	1	1	1	1	1	1
0	0	0	0	0	1	1	1	1	1	1	1	1	1	0	1	1	1	1	1	1	1
0	0	1	0	0	1	1	1	1	1	1	1	1	1	1	0	1	1	1	1	1	1
0	0	0	1	0	1	1	1	1	1	1	1	1	1	1	1	0	1	1	1	1	1
0	0	1	1	0	1	1	1	1	1	1	1	1	1	1	1	1	0	1	1	1	1
0	0	0	0	1	1	1	1	1	1	1	1	1	1	1	1	1	1	0	1	1	1
0	0	1	0	1	1	1	1	1	1	1	1	1	1	1	1	1	1	1	0	1	1
0	0	0	1	1	1	1	1	1	1	1	1	1	1	1	1	1	1	1	1	0	1
0	0	1	1	1	1	1	1	1	1	1	1	1	1	1	1	1	1	1	1	1	0

(B) Truth table.

Fig. 10-13.　A 1-out-of-16 decoder.

218

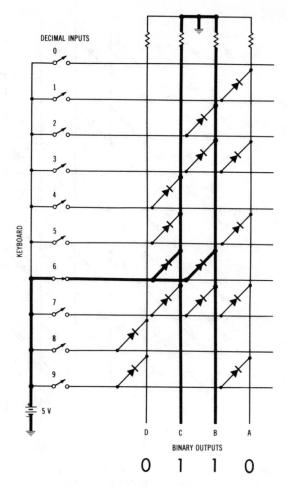

Fig. 10-14. Decimal-binary conversion with diode matrix.

DECIMAL–SEVEN-SEGMENT MATRIX

Conversion from decimal to seven-segment code may be accomplished by a diode-matrix ROM as shown in Fig. 10-15. Ten input lines and seven output lines are interconnected by means of diodes. For example, if the "7" digit-select line is driven logic-high, segments a, b, and c in the display device are activated, and the decimal digit "7" is displayed. (Refer again to Figs. 10-11 and 10-12.)

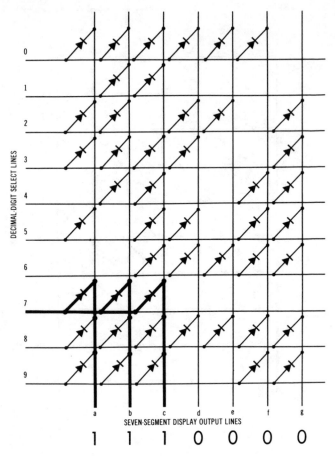

Fig. 10-15. Diode matrix for conversion from decimal to
seven-segment code.

ADDRESS DECODER

An address is an expression, such as a binary number, that
designates a specific location in a storage or memory device
(such as a ROM). An address may also identify some other
source or destination of data in a computer. An address must be
decoded into a machine language (computer code). Machine
language comprises a set of symbols, characters, or signs and
the rules for combining them in order to convey instructions or
data to a computer. An example of how a binary address is decoded into machine language is shown in Fig. 10-16A.

The binary address inputs for the X and Y decoders are $A_0A_1A_2$ and $A_3A_4A_5$, respectively. The decoded address outputs are $X_0X_1X_2X_3X_4X_5X_6X_7$ and $Y_0Y_1Y_2Y_3Y_4Y_5Y_6Y_7$, respectively. As an illustration, an inputted binary address might be:

$$A_0A_1A_2A_3A_4A_5 = 110010$$

The truth table for the X decoder (Fig. 10-16B) is:

A_0	A_1	A_2	High Output
0	0	0	X_0
0	0	1	X_1
0	1	0	X_2
0	1	1	X_3
1	0	0	X_4
1	0	1	X_5
1	1	0	X_6
1	1	1	X_7

The Y decoder operates similarly.

Since 110 corresponds to decimal 6, or to decoder output X_6, this line goes high. Since 010 corresponds to decimal 2, or to decoder output Y_2, this line goes high. In other words, memory cell X_6Y_2 is accessed by the inputted binary address 110010.

Observe that, in practice, four-input AND gates would ordinarily be used in Fig. 10-16B. The fourth input is called an enable input; the enable input is employed to disable the gate(s) when desired. That is, whenever an enable input is logic-low, the gate output will remain low regardless of the address-input states. For example, with reference to the truth table, X_0 will be high when $A_0A_1A_2 = 000$. However, it the X_0 gate has an enable input and this input is driven logic-low, then X_0 will be logic-low no matter what the value of $A_0A_1A_2$ may be.

HEXADECIMAL-BINARY DECODER

The hexadecimal system has a base of 16 and employs the digits 0 through 9 and six other symbols, typically A through F. Programmers use the hexadecimal system because it is more compact than the binary system or the octal system; hexadecimal numbers are comparatively easy to read and comparatively easy to remember. Equivalent hexadecimal and decimal numbers are listed in Table 10-1.

Unless an operator has had considerable experience with

221

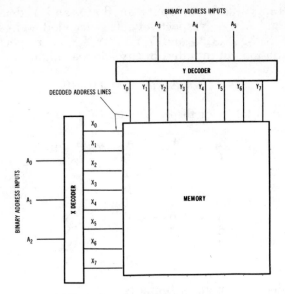

(A) Decoder connections to memory.

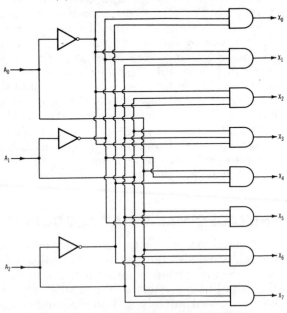

(B) Logic diagram of 1-of-8 decoder.

Fig. 10-16. Arrangement of an address decoder.

Table 10-1. Hexadecimal and Decimal Numbers

Hexadecimal Number	Decimal Number
0	0
1	1
2	2
3	3
4	4
5	5
6	6
7	7
8	8
9	9
A	10
B	11
C	12
D	13
E	14
F	15

hexadecimal arithmetic, calculations will be facilitated by changing the literal symbols into their decimal equivalents. For example, the hexadecimal number 2AD may be written in the form 2 10 13 for purposes of addition. Note that 2AD has the following decimal value:

$$2\times16^2 + 10\times16^1 + 13\times16^0 = 685_{10}$$

The hexadecimal number 3CF may be written in the form 3 12 15; it in turn has the decimal value:

$$3\times16^2 + 12\times16^1 + 15\times16^0 = 975_{10}$$

Of course, the sum of 685_{10} and 975_{10} is 1660_{10}. The hexadecimal addition can be performed as follows:

$$
\begin{array}{rccc}
2AD = & 2 & 10 & 13 \\
+3CF = & 3 & 12 & 15 \\
\hline
& 5 & 22 & 28
\end{array}
$$

Note that the sums of the last two columns are numbers larger than 15, so they cannot be represented in hexadecimal notation as they are. This problem is dealt with as follows: Because these are hexadecimal numbers, the number 28 in the right col-

223

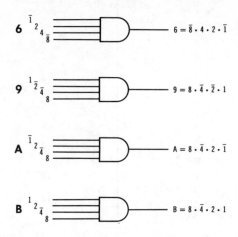

Fig. 10-17. Hexadecimal (binary format) encoder.

umn can be reduced by 16 and 1 carried to the next column to the left. The sum of this column becomes 23:

$$
\begin{array}{ccc}
 & 1 & \\
2 & 10 & 13 \\
3 & 12 & 15 \\
\hline
5 & 23 & 12 \\
\end{array}
$$

Similarly, the 23 can be reduced by 16 and 1 carried to the next column:

$$
\begin{array}{ccc}
1 & 1 & \\
2 & 10 & 13 \\
3 & 12 & 15 \\
\hline
6 & 7 & 12 \\
\end{array}
$$

The result is thus hexadecimal 67C, which is equivalent to decimal 1660.

As in hexadecimal addition, it is usually helpful to change the literal symbols into their decimal equivalents to perform hexadecimal subtraction. If a borrow is required during a subtraction operation, 16 is added to the desired number, and the digit borrowed from is decreased by 1. Hexadecimal multiplication is cumbersome because a hexadecimal multiplication table has 256 entries. Therefore, programmers generally convert the multiplier and the multiplicand into their decimal equivalents, perform the multiplication, and then convert the product into its hexadecimal equivalent. Similarly, hexadecimal division is cumbersome, and programmers generally convert the divisor and the dividend into the decimal equivalents, perform the division, and then convert the quotient into its hexadecimal equivalent.

Hexadecimal characters expressed in binary format are decoded by means of AND gates, as shown in Fig. 10-17. The number 69AB expressed in binary format is written:

$$
\begin{array}{cccc}
6 & 9 & A & B \\
0110 & 1001 & 1010 & 1011 \\
\text{or,} \ \ \overline{8}42\overline{1} & 8\overline{4}21 & 8\overline{4}2\overline{1} & 8\overline{4}21
\end{array}
$$

The binary format is processed in groups of four bits called *nibbles*. Each nibble represents a hexadecimal character. There are 16 hexadecimal characters 0 through F. In turn, one 4-input AND gate is utilized to convert each nibble into a hexadecimal output.

Note the following relations:

$$
\begin{array}{cccc}
6 & 9 & A & B \\
16^3 & 16^2 & 16^1 & 16^0 \\
\times 6 & \times 9 & \times 10 & \times 11 \\
\hline
24576 & 2304 & 160 & 11
\end{array}
$$

$$24576 + 2304 + 160 + 11 = 27051$$

That is, $\overline{8}42\overline{1} \ 8\overline{4}2\overline{1} \ 8\overline{4}2\overline{1} \ 8\overline{4}21 = 27{,}051_{10}$.

Chapter
11

Parity Generator/Checkers
and Interfacing

This chapter deals with two important aspects of the practical use of digital circuits: parity checking and interfacing. Parity checking is a method used to detect errors in the processing or transmission of digital signals. Interfacing has to do with the transfer of digital information between points that are not directly compatible.

PARITY CHECKING AND
ERROR CORRECTION

Since digital systems are not completely reliable, a basic requirement concerns recognition of processing errors in digital data. For example, if a particular bit is incorrectly read as a 0 instead of a 1 in a digital word, an erroneous answer would be obtained. Therefore, a suitable method of checking and detecting an erroneous bit is required.

A widely used method of detecting errors in digital words is to count the number of logic-1 bits in a 7-bit character, as follows:

Character or Command	Code
A	1000001
B	1000010
C	1000011

An even number of logic-1 bits is present in the digital code word for A and for B; an odd number of logic-1 bits is present in the digital code word for C. When the number of logic-1 bits is odd, one more logic-1 bit is added to the group to form an 8-bit word that contains an even number of logic-1 bits. On the other hand, when the number of logic-1 bits in the 7-bit character is even, a logic-0 bit is added in the eighth position, so that the 8-bit digital code word contains an even number of logic-1 bits. Thus:

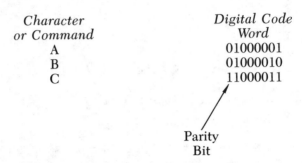

Character or Command	Digital Code Word
A	01000001
B	01000010
C	11000011

Parity
Bit

Accordingly, every 8-bit digital word contains an even number of logic-1 bits. When these 8-bit words are transferred into or out of a memory, a receiving device counts the logic 1s in each word to verify that the number is even. An odd number of logic-1 bits in a word causes an error signal to be triggered. This process is called *parity checking*, and the particular method that has been described is termed *even parity*. Odd parity checking is also employed in digital systems; in that case, each valid binary word normally contains an odd number of logic-1 bits.

Although a system malfunction could cause two logic-1 bits in the same 8-bit digital word to drop out, this is not probable from a statistical viewpoint. Of course, if a double drop-out should occur in the same binary word, the parity check would pass the word as valid. By adding more bits to each binary word, and by elaborating the encoding and detection network, error checking processes can be made more reliable. A sophisticated arrangement can detect which bit in a binary word is erroneous. In addition, the system can be designed so that the erroneous bit is automatically corrected; these functions are commonly provided in large computing systems.

A *parity generator/checker* is used to generate a parity bit for transmission with a data word, or to check a data word for odd parity or even parity. An example of a nine-bit parity generator/checker is shown in Fig. 11-1. In order to generate a parity bit for transmission with a data word, the data word is inputted on

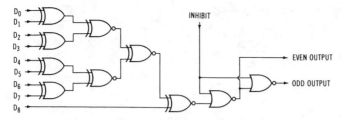

Fig. 11-1. A nine-bit parity generator/checker.

lines D_0 through D_7. If even parity is employed, a logic 0 is supplied on line D_8. If odd parity is utilized, a logic 1 is supplied on line D_8. In odd-parity operation (line D_8 high), the odd output line will be high only when it is necessary to add a logic-1 parity bit in the ninth bit place (nine-bit word including parity bit). This parity bit will be transmitted along with the data word as the ninth bit.

At the receiving end, the eight-bit digital word with a ninth parity bit is inputted to the nine inputs of a similar parity checker/generator. Inasmuch as the parity of the digital words is known when the data is transmitted, any change of this parity that is detected during the receiving-end check indicates a transmission error. For example, suppose that odd parity is employed; then, an indication of even parity (logic 1 on the even output) would indicate that an error had occurred.

Fig. 11-2 shows examples of *parity trees*, which are so named for their resemblance to trees with various branches. Each is a group of XNOR gates that can be used to check a number of input bits for either odd or even parity. Parity trees are used both to check and to generate parity wherever a redundant bit is added to a digital word in order to check for error.

Parity Generator/Checker IC Packages

An example of an eight-bit parity generator/checker IC package is shown in Fig. 11-3. Eight data inputs are provided. There are also an odd parity input and an even parity input; these parity inputs control the output states as indicated in the truth table. For example, if the even-parity input is held high and the odd-parity input is held low, the sum-even output (ΣQ_E) will go high if the number of 1s at data inputs 0 through 7 is even; otherwise, the sum-even output will go low. In all parity-check output states, the sum-even and sum-odd lines are complementary.

A nine-bit parity generator/checker IC package is shown in Fig. 11-4. Parity-even and parity-odd outputs are provided, with

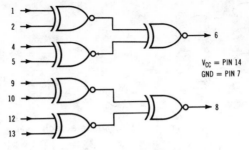

(A) Dual four-bit parity tree.

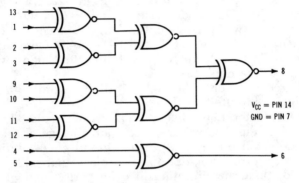

(B) Eight-bit parity tree.

Fig. 11-2. Examples of parity trees.

an enable line. This arrangement is employed with eight-bit digital words plus a parity bit. The even-parity output (PE) will be logic-high if an even number of logic 1s is present on the nine inputs. On the other hand, the odd-parity output (PO) will be logic-high if an odd number of logic 1s is present on the nine inputs. Note that the enable input ($\overline{E}$) forces both of the outputs low when it is driven high (the enable input is active-low).

A 12-bit parity generator/checker IC package is shown in Fig. 11-5. Although parity-even and parity-odd outputs are provided, no inhibit or enable line is included. As a digital device becomes more complex, the pinout requirements increase also. In turn, various tradeoffs are made to retain a reasonably sized package. This is an example of medium-scale integration (MSI), whereas a quad 2-input NAND gate is an example of small-scale integration (SSI). This 12-bit parity generator/checker IC package employs 16 pins; if an inhibit or enable input were included, the next larger MSI package (24 pins) would be required.

230

Troubleshooting Notes

Parity generator/checkers are widely utilized in data-bus arrangements. A *bus* (Fig. 11-6) is simply one or many conductors (such as printed-circuit conductors) used as a path over which digital information is transmitted from any of several sources to any of several destinations. A *data bus* carries digital data to or from a number of different locations. The term *data* denotes basic elements of information that can be processed or produced by a digital computer. A fundamental configuration employing odd-parity generator/checkers is shown in Fig. 11-7. The parity bits for even and odd parity are shown in Table 11-1.

Short circuits occasionally occur in buses. For example, there are many troubleshooting situations wherein a faulty circuit node is found to be "stuck," and if there are many elements common to this node, a "tough-dog" problem can result. For example, in Fig. 11-8, line A is shorted to the supply voltage, and line B is shorted to ground. Any signal line connected to A will be "stuck high," and any signal line connected to B will be "stuck low."

The "stuck-at" type of malfunction can be analyzed quickly and nondestructively by using a digital-logic current tracer. The advantage of this approach is that it is not necessary to "lift" IC pins, cut PC conductors, or force large amounts of current into the conductor in an attempt to burn out the short. Note that a current tracer held over a current path shows whether or not pulses of current are present. In some cases, lack of current is due to a dead node driver. Most important, a current tracer shows *where* a pulse current is flowing. Observe that if a node is stuck low and the reason for the fault is a shorted input on one of the components of the node, there will be a heavy pulse current between the node driver and the faulty component. A logic pulser is often used with a current tracer as illustrated in Fig. 11-9; in this example, a solder bridge is pinpointed.

INTERFACES

Many types of interfaces are encountered in digital logic circuits. An interface is a point or a device at which a transition between media, power levels, modes of operation, or related functions is made. One of the simplest interfaces is an MOS/bipolar or bipolar/MOS interface to translate the differing voltage levels between MOS circuitry and bipolar circuitry (Fig. 11-10). An analog-to-digital converter or a digital-to-analog converter is a more complex example of interfacing. A simple interface may be integral with another device, as in the example of Fig. 11-11. This is a dual high-voltage/high-current driver in NAND-gate form. It employs circuitry that can withstand 30 V in the high state and that can sink 500 mA in the low state. The inputs are TTL compatible and have clamp diodes for transient suppression. A strobe input is provided to eliminate race problems, and an expander input is available on each gate for

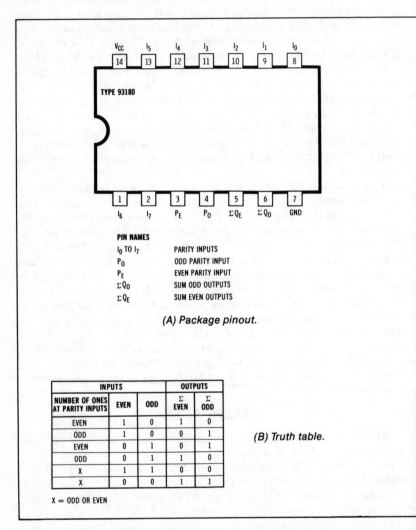

(A) Package pinout.

INPUTS			OUTPUTS	
NUMBER OF ONES AT PARITY INPUTS	EVEN	ODD	Σ EVEN	Σ ODD
EVEN	1	0	1	0
ODD	1	0	0	1
EVEN	0	1	0	1
ODD	0	1	1	0
X	1	1	0	0
X	0	0	1	1

X = ODD OR EVEN

(B) Truth table.

Fig. 11-3. An eight-bit parity generator/checker IC package.

232

input-diode expansion. Separate ground pins are provided for each gate to minimize ground-current disturbances at high-current operating levels. Typical load circuitry might include indicator LEDs and switching diodes for the selection of oscillator crystals.

An example of an interface for TTL/MOS operation is shown in Fig. 11-12. Voltage levels are translated; most MOS circuitry

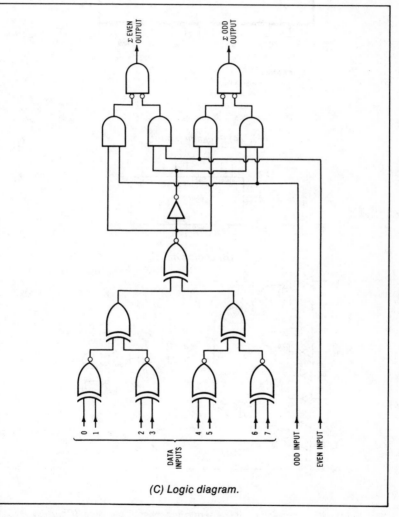

(C) Logic diagram.

(Courtesy Fairchild Camera and Instrument Corp.)

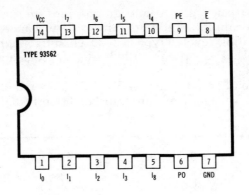

PIN NAMES

I_0 TO I_8	DATA INPUTS
$\overline{E}$	OUTPUT ENABLE
PO	ODD PARITY OUTPUT
PE	EVEN PARITY OUTPUT

(A) Package pinout.

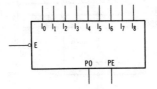

(B) Logic symbol.

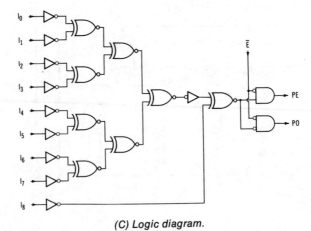

(C) Logic diagram.

Fig. 11-4. A nine-bit parity generator/checker IC package. (*Courtesy Fairchild Camera and Instrument Corp.*)

234

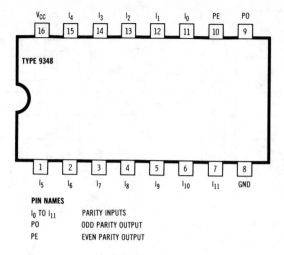

PIN NAMES

I_0 TO I_{11}	PARITY INPUTS
PO	ODD PARITY OUTPUT
PE	EVEN PARITY OUTPUT

(A) Package pinout.

(B) Truth table.

INPUTS	OUTPUTS	
	PO	PE
ODD NUMBER HIGH	1	0
EVEN NUMBER HIGH	0	1

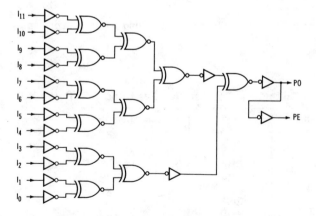

(C) Logic diagram.

Fig. 11-5. A 12-bit parity generator/checker IC package. (*Courtesy Fairchild Camera and Instrument Corp.*)

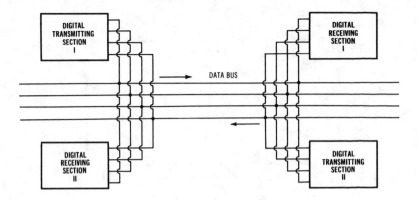

Fig. 11-6.   A bidirectional data bus.

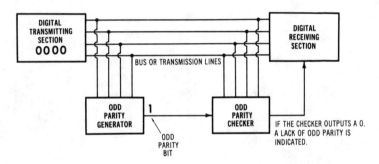

Fig. 11-7.   A four-bit data-transmitting system with odd parity.

Table 11-1. Comparative Even and Odd Parity Bits

Decimal	BCD	Even Parity Bit	Odd Parity Bit
0	0000	0	1
1	0001	1	0
2	0010	1	0
3	0011	0	1
4	0100	1	0
5	0101	0	1
6	0110	0	1
7	0111	1	0
8	1000	1	0
9	1001	0	1

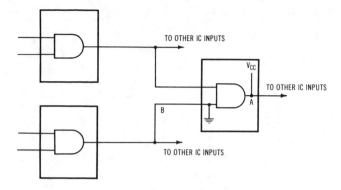

Fig. 11-8. Examples of short circuits in an IC. (*Courtesy Hewlett-Packard*)

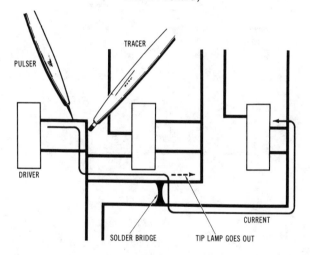

Fig. 11-9. Method of finding solder bridge. (*Courtesy Hewlett-Packard*)

uses negative power supplies and voltage swings in the range of −5 to −30 V. On the other hand, TTL circuitry employs positive power supplies with voltage swings in the range from 0 to 5 V. The interface in Fig. 11-12 is called a hex inverter buffer/driver. It features open-collector output that can operate at comparatively high voltage for working into MOS circuitry. Note also that this interface may be used to drive comparatively high-current loads such as lamps or relays. When MOS circuitry works into TTL circuitry, a reverse interface arrangement is utilized, as was shown in Fig. 11-10; in that example, both V_{CC} and V_{DD} power-supply pins were provided.

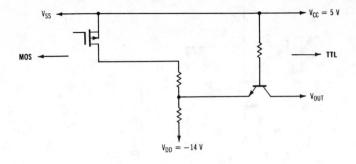

(A) Diagram of basic circuit.

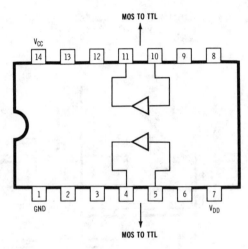

(B) Example of IC package.

Fig. 11-10. Interfacing between MOS and TTL.

POSITIVE LOGIC, NEGATIVE LOGIC, POSITIVE OPERATING LEVEL, NEGATIVE OPERATING LEVEL

It was mentioned previously that both positive operating levels and negative operating levels may be encountered in digital logic circuits. Moreover, both positive logic and negative logic are in general use. As Fig. 11-13 shows, operation between +5 V and ground is an example of a positive operating level. When +5 V represents logic 1 and ground represents logic 0, positive logic is being used. Conversely, if +5 V repre-

238

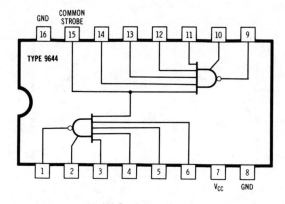

(A) Package pinout.

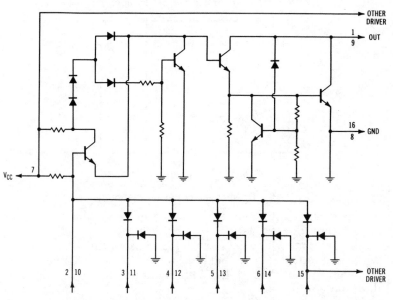

(B) Logic diagram (one section).

Fig. 11-11. A dual high-voltage, high-current driver IC.

sents logic 0 and ground represents logic 1, negative logic is being used. When operation takes place between -12 V and ground, a negative operating level is employed. In this case, if ground represents logic 0 and -12 V represents logic 1, positive logic is being used. On the other hand, if ground represents logic 1 and -12 V represents logic 0, negative logic is being utilized. (Note that a logic circuit remains physically unchanged

239

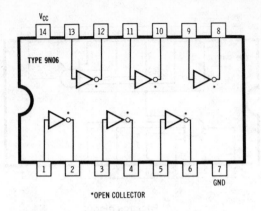

(A) Pinout diagram of package.

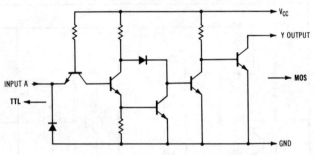

(B) Schematic diagram (one inverter).

Fig. 11-12.   Open-collector hex inverter buffer/driver.

whether its circuit action is discussed in terms of positive logic or negative logic.)

Observe that when negative logic is used, the 1s and 0s that we are familiar with in positive logic become changed (complemented) into 0s and 1s. With reference to Fig. 11-14, note the following facts:

1. An AND gate in a positive-logic circuit will "look like" an OR gate when operated in a negative-logic circuit. In other words, if the 1s and 0s in the AND-gate truth table are changed into 0s and 1s, we obtain the OR-gate truth table.

2. An OR gate in a positive-logic circuit will "look like" an AND gate when operated in a negative-logic circuit. That is, if we change the 1s and 0s in the OR-gate truth table into 0s and 1s, we obtain the AND-gate truth table.

3. A NAND gate in a positive-logic circuit will "look like" a NOR gate when operated in a negative-logic circuit.

240

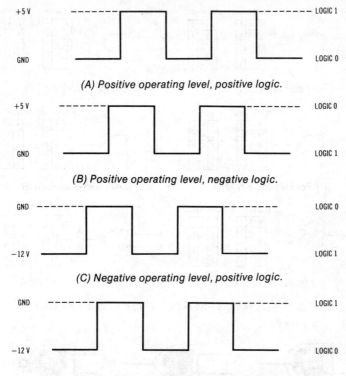

(A) Positive operating level, positive logic.

(B) Positive operating level, negative logic.

(C) Negative operating level, positive logic.

(D) Negative operating level, negative logic.

Fig. 11-13. Examples of positive and negative operating levels with positive and negative logic conventions.

4. A NOR gate in a positive-logic circuit will "look like" a NAND gate when operated in a negative-logic circuit.
5. An XOR gate in a positive-logic circuit will "look like" an XNOR gate when operated in a negative-logic circuit.
6. An XNOR gate in a positive-logic circuit will "look like" an XOR gate when operated in a negative-logic circuit.

These facts are summarized in Fig. 11-15.

Similarly, an AND-OR combination of gates in a positive-logic circuit will "look like" an OR-AND combination of gates when operated in a negative-logic circuit. In other words, each AND gate will "look like" an OR gate, each OR gate will "look like" an AND gate, and so on, as shown in the example in Fig. 11-16.

If a digital system or a section of a digital system is operating with negative logic, it may be analyzed as follows:

241

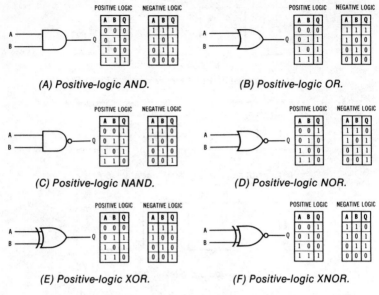

(A) Positive-logic AND. *(B) Positive-logic OR.*

(C) Positive-logic NAND. *(D) Positive-logic NOR.*

(E) Positive-logic XOR. *(F) Positive-logic XNOR.*

Fig. 11-14. Positive and negative logic for common gates.

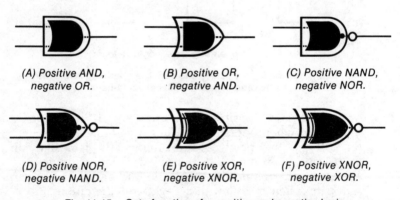

(A) Positive AND, negative OR.

(B) Positive OR, negative AND.

(C) Positive NAND, negative NOR.

(D) Positive NOR, negative NAND.

(E) Positive XOR, negative XNOR.

(F) Positive XNOR, negative XOR.

Fig. 11-15. Gate functions for positive and negative logic.

1. Change all of the AND gates to OR gates; change all of the OR gates to AND gates; change all of the NAND gates to NOR gates; change all of the NOR gates to NAND gates; change all of the XOR gates to XNOR gates; change all of the XNOR gates to XOR gates.
2. In case a subsection in the negative-logic area is described in terms of its truth table, complement all terms in the

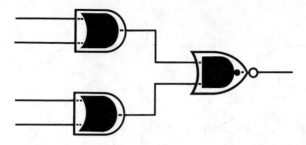

Fig. 11-16. Example of a positive-logic circuit and its "appearance" with negative logic.

truth table; that is, change all 1s to 0s, and change all 0s to 1s.
3. Analyze the resulting system as if it were a positive-logic system.

RS-232C INTERFACE FUNCTION

The designation RS-232C identifies an EIA standard that defines the interfacing between data-terminal equipment and data-communication equipment for serial binary data interchange over wire lines. *Mark* and *space* signals, which correspond to logic-high and logic-low levels, are employed. A mark potential is at least −3 volts; a space potential is at least +3 volts. An RS-232C line driver functions to convert TTL logic levels to EIA levels for transmission between data-terminal equipment and data-communication equipment. Conversely, an RS-232C line receiver functions to convert EIA signal levels to TTL logic levels. An example of a triple RS-232C line-driver IC is shown in Fig. 11-17. Each driver consists of an AND-OR-invert configuration for positive or negative voltage output.

An RS-232C line receiver is essentially a level-shifting inverter; it inputs signals from an RS-232C driver and outputs TTL-compatible logic levels. An example of a triple RS-232C line-receiver IC is shown in Fig. 11-18. Each channel has three inputs: conventional buffer input, hysteresis (slicing) input, and response (control) input. Fig. 11-18B shows the details of the configuration. Time slices may be regarded as time quanta of a few hundred milliseconds; they are employed in time-shared data communication systems. The response input is associated with a data communication control technique ("handshaking") based on data transfer requests and ready and acknowledge signals.

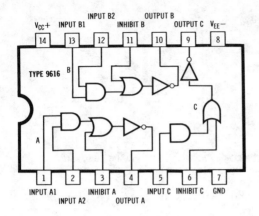

INPUT			OUTPUT
1	2	INHIBIT	
0	0*	0	1
1	1*	0	0
0	0*	1	0
1	1*	1	0
0*	1*	0*	1*
1*	0*	0*	1*
0*	1*	1*	0*
1*	0*	1*	0*

*CHANNELS A AND B ONLY

(A) Package pinout.   *(B) Truth table.*

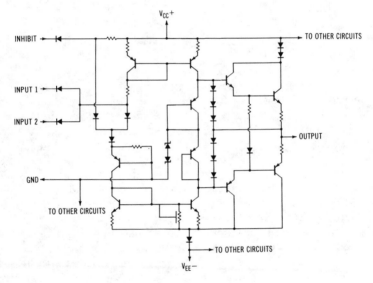

(C) Circuit of one channel.

Fig. 11-17.   A triple RS-232C line-driver IC. (*Courtesy Fairchild Camera and Instrument Corp.*)

DIFFERENTIAL LINE DRIVER

Another common type of interfacing in line systems employs a differential line driver to convert a single-ended source voltage into a double-ended line voltage. The advantage of the push-pull (balanced) mode of line operation is minimization of

244

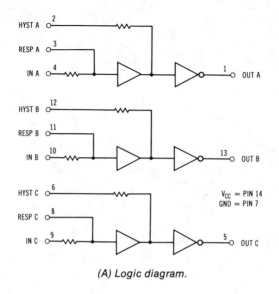

(A) Logic diagram.

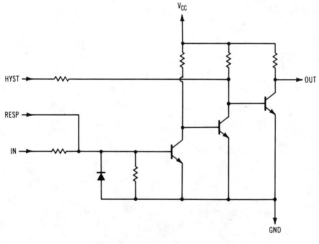

(B) Circuit of one channel.

Fig. 11-18. A triple RS-232C line-receiver IC. (*Courtesy Fairchild Camera and Instrument Corp.*)

interference due to common-mode noise. The dual differential line driver shown in Fig. 11-19 provides three-state push-pull output, and it may also be operated with a single-ended three-state output if desired. When differential line drive is used, a differential line receiver (Fig. 11-20) is also utilized. It inputs

245

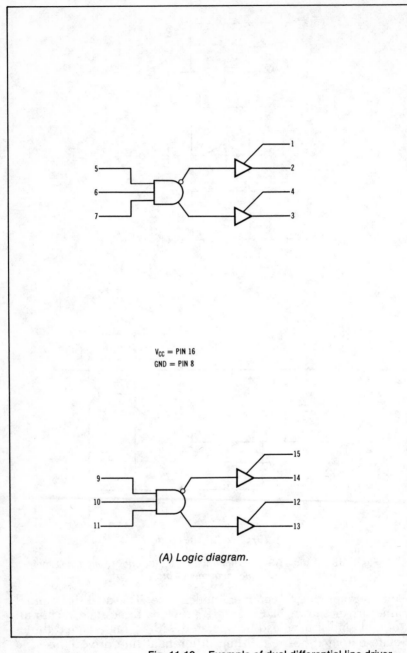

V_CC = PIN 16
GND = PIN 8

(A) Logic diagram.

Fig. 11-19.   Example of dual differential line driver.

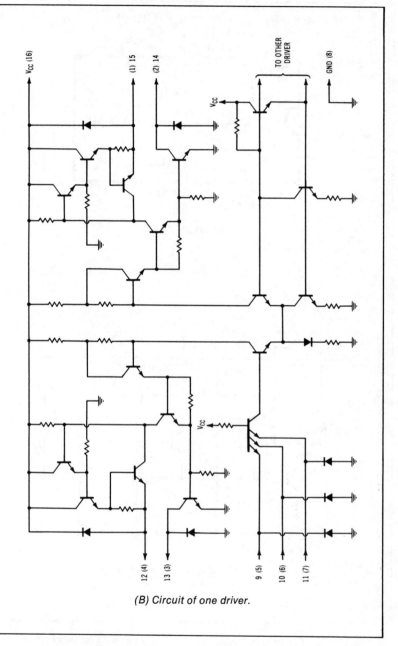

(B) Circuit of one driver.

(Courtesy Fairchild Camera and Instrument Corp.)

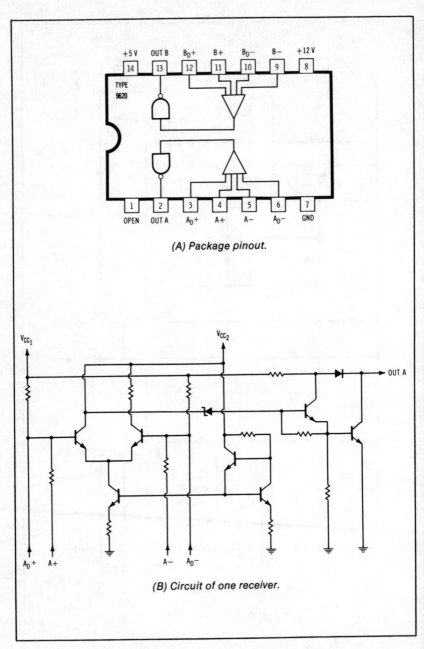

Fig. 11-20. Example of dual differential line receiver. (*Courtesy Fairchild Camera and Instrument Corp.*)

248

push-pull digital signals, rejects common-mode interference, and outputs standard TTL single-ended pulses. Note that both direct and attenuated input terminals are provided; attenuated input is ordinarily employed so that external components can be included for control of response time and attenuation level. This interface arrangement provides open-collector output.

THREE-STATE BUS DRIVER

A three-state bus driver is similar to a three-state buffer, which was described in Chapter 2. The three-state mode of operation finds wide application in bus-oriented digital systems, because a bus often carries digital data in either direction to various destinations. A driver at one end of a bus may be enabled and then turned off so that a driver at the other end of the

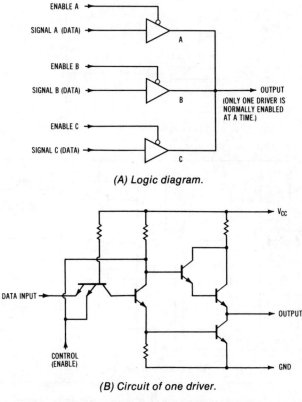

(A) Logic diagram.

(B) Circuit of one driver.

Fig. 11-21. Example of a triple three-state bus driver.

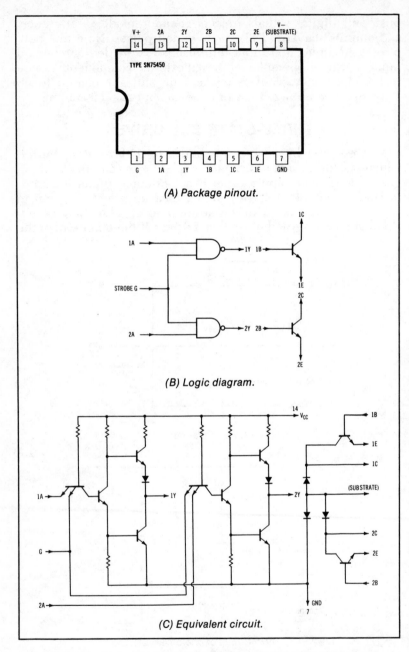

Fig. 11-22. A dual peripheral driver. (*Courtesy Fairchild Camera and Instrument Corp.*)

bus may be enabled. Or, a driver in the middle of the bus might be enabled. A three-state driver has three output states: high, low, and floating (open). In its floating state, the driver has a very high output impedance (practically an open circuit). A three-state bus-driver arrangement is shown in Fig. 11-21. As can be seen in Fig. 11-21B, driving the enable input low serves to cut off both of the totem-pole transistors, thereby providing a "floating" output.

PERIPHERAL DRIVER

Another type of interface, called a *peripheral driver*, is frequently encountered. An example of a peripheral-driver IC is shown in Fig. 11-22. It consists of two NAND gates with totem-pole outputs, and two high-current high-voltage driver transistors with open collectors. Peripheral drivers are used to energize relays, lamps, lines, and other loads that require appreciable power. A clock line in an elaborate digital system, for example, has a large number of loads; therefore, the clock driver must supply a corresponding amount of power.

TROUBLESHOOTING TECHNIQUES

System malfunction is occasionally caused by a short from V_{CC} to ground external to the IC packages. To pinpoint the short, a logic pulser and current tracer may be used as shown in Fig. 11-23. It is advisable to lift one side of the electrolytic

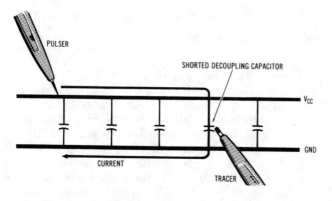

Fig. 11-23. Method of finding short from voltage source to ground.
(*Courtesy Hewlett-Packard*)

Table 11-2. Bus (Line) Troubleshooting Basics

TYPE OF BUS DRIVER	TROUBLESHOOTING TIPS
Open Collector (Wire-AND/OR)	1. Open collectors can sink current but not source it, so a pull-up resistor to V_{CC} is connected to the output. 2. Disable driver input(s). 3. Pulse output(s). 4. Faulty driver will draw the most current.
Single Driver	1. Driver can both source and sink current. 2. Pulse input(s). 3. Probe output(s) for logic state changes, **OR** 4. Current trace output(s) for amplitude and the direction of the current path. 5. Determine if driver is dead or bus is stuck. 6. Replace dead driver, **OR** 7. Pulse and current trace at output to pinpoint bus fault.
Three-State Buffer (With Source and Sink Capability)	1. Disable driver inputs. 2. Pulse bus output lines. 3. If one output draws current, verify if it is faulty **OR** 1. Enable drivers. 2. Pulse driver inputs individually. 3. If one output fails to indicate current, verify if it is open.

Courtesy Hewlett-Packard

capacitors on the V_{CC} line because electrolytics "eat" test pulses and produce confusing current paths. Inject test pulses across the power-supply pins or across components in the corners of the pc board. Move the pulsing point from corner to corner, and trace the current from the pulsing point with the current tracer. Because the test pulses are flowing into a short circuit, the current indication is strong; the indicator lamp in the current tracer glows in proportion to the current intensity at its tip. As the tip is moved past the short-circuit point, the indicator lamp will go dark. When it appears that the short has been located, move the pulse-injection point to the short. If this is the true short point, no current paths will be detected elsewhere on the pc board.

A bus can become "stuck" because of an open circuit at the bus-driver output, in the bus line itself, or at the input of a data receiver. Usually, the open circuit can be located by pulsing and probing or by using a current tracer to determine whether or not current is present at inputs or outputs. Some tips for bus troubleshooting are listed in Table 11-2.

Chapter 12

Code Converters, Multiplexers, Demultiplexers, and Comparators

There are a number of different codes that may be used in digital circuits, and the need often arises to change from one code to another. This chapter will deal with code conversions and will also discuss the processes of multiplexing and comparison.

CODE CONVERTERS

A code converter is a combinatorial logic arrangement that changes coded digital data from one format into another. From a fundamental viewpoint, a code converter can be regarded as an interface. As an illustration, a digital system may include means for converting the 8421 binary code into the 2'421 binary code, or vice versa. Both of these codes were noted previously. The code converter is based on the truth tables for the two codes, as shown in Fig. 12-1. Note the following relations between the two truth tables:

Numbers 0 through 4 have the same coded forms.
Numbers 5 through 9 have different coded forms.
The 1 bit is identical in both codes.

It follows that we can use the 1 bit of the 2'421 code to generate the 1 bit of the 8421 code.

The 2 bit in the 8421 code equals the 2 bit in the 2'421 code when 2' is NOT true (logic-low).

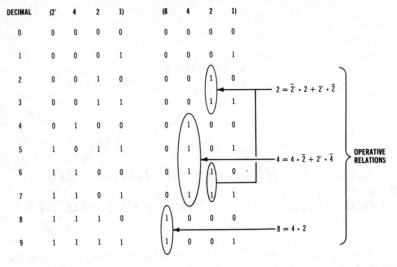

Fig. 12-1. Comparison of truth tables for 2'421 and 8421 bcd codes.

When 2' is true (logic-high), the complement of the 2 bit in the 2'421 code is used by the converter.

From these two facts, we can write the following logic equation:

$$2 = (\overline{2'} \cdot 2) + (2' \cdot \overline{2}) = 2' \oplus 2$$

We also observe from the truth tables that the 4 bit in the 8421 code occurs (is logic-high) whenever the 4 bit in the 2'421 code is true (logic-high) and the 2 bit in the 2'421 code is NOT true (logic-low). Also, the 4 bit for the fifth count can be generated by 2' AND $\overline{4}$, inasmuch as this combination does not occur on any other count. These facts lead to the logic equation:

$$4 = (4 \cdot \overline{2}) + (2' \cdot \overline{4})$$

Finally, the 8 bit in the 8421 code can be generated from 4 and 2 in the 2'421 code, since $(4 \cdot 2)$ is NOT true (logic-low) in the first seven counts.

These logic relations are typically implemented by means of the arrangement of five gates in Fig. 12-2. Other binary codes can be converted in the same general manner.

256

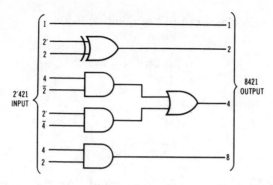

Fig. 12-2. Logic circuit for converting 2'421 code to 8421 code.

Octal Code Conversion

The octal number system has a base of 8 and employs the digits 0 through 7. An advantage of the octal system is that it is more compact and the octal numbers are easier to read and easier to remember than are binary numbers. A comparison of the first 13 octal, binary, and decimal numbers is shown in Table 12-1.

Table 12-1. Decimal, Binary, and Octal Equivalents

Decimal	Octal	Binary	Decimal	Octal	Binary
0	0	0	7	7	111
1	1	1	8	10	1000
2	2	10	9	11	1001
3	3	11	10	12	1010
4	4	100	11	13	1011
5	5	101	12	14	1100
6	6	110	13	15	1101

Octal addition obeys the following rules:

1. If the sum of any column is equal to or greater than the base (equal to or greater than 8), then 8 must be subtracted from the sum to obtain the final result of the column.
2. If the sum of any column is equal to or greater than the base (equal to or greater than 8), there will be a carry; the value of the carry will be equal to the number of times that the base value was subtracted.
3. If the result of any column addition is less than the value of the base (less than 8), the base value is not subtracted, and no carry is generated.

257

Note the following example of octal addition:

	Octal			Decimal
		35		29
		+63		+51
	1	10	8	80
	−8	−8		
	1	2	0	

$$120_8 = 80_{10}$$

Octal subtraction can be performed directly, as in decimal subtraction. Whenever a borrow is needed, an 8 is borrowed and added to the number. Note the following example of octal subtraction:

$$\begin{array}{r} 2022_8 \\ -\,1234_8 \\ \hline 566_8 \end{array}$$

Octal multiplication is performed with the aid of an octal multiplication table, much as decimal multiplication is performed with the aid of a decimal multiplication table. All additions required in the multiplication procedure obey the rules for octal addition. An octal multiplication table is shown in Table 12-2. Note the following example of octal multiplication:

	Octal				Decimal
		177			127
		×27			×23
		1571			381
		376			254
5	13	13	1		2921
−0	−8	−8	−0		
5	5	5	1		

$$5551_8 = 2921_{10}$$

Octal numbers are multiplied by looking up the product in the table. Any product equal to or larger than the base or radix (larger than 8) results in a carry, which is then added octally to the next product. The final product is then summed up by octal addition.

Octal division is performed in much the same manner as dec-

Table 12-2. Octal Multiplication Table

×	0	1	2	3	4	5	6	7
0	0	0	0	0	0	0	0	0
1	0	1	2	3	4	5	6	7
2	0	2	4	6	10	12	14	16
3	0	3	6	11	14	17	22	25
4	0	4	10	14	20	24	30	34
5	0	5	12	17	24	31	36	43
6	0	6	14	22	30	36	44	52
7	0	7	16	25	34	43	52	61

imal division. However, all of the multiplication and subtraction operations are done in octal format. Note the following example of octal division:

$$
\begin{array}{r}
\textit{Octal} \\
66 \\
22 \overline{) \, 1714} \\
154 \\
\hline
154 \\
154 \\
\hline
000
\end{array}
\qquad
\begin{array}{r}
\textit{Decimal} \\
54 \\
18 \overline{) \, 972} \\
90 \\
\hline
72 \\
72 \\
\hline
00
\end{array}
$$

$$66_8 = 54_{10}$$

Binary-to-octal, decimal-to-octal, hexadecimal-to-octal, octal-to-binary, octal-to-hexadecimal, and octal-to-decimal converters are easily implemented by means of diode logic in matrix form, as described previously for binary-to-decimal and decimal-to-binary conversion.

Code Conversion in Digital Slot Machine

A simple digital slot machine provides an example of a binary-to-alphameric code converter. (Alphameric is a generic term for alphabetic letters, numerical digits, and special characters.) The logic circuitry for one "reel" of a digital slot machine is shown in Fig. 12-3. Other "reels" duplicate the configuration; each "reel" has an independent clock. The logic circuitry functions to convert the binary sequence from the flip-flops into seven-segment characters representing "cherry," "lemon," "orange," and "apple." Thus, each logic section is a binary-to-alphameric code converter.

With reference to Fig. 12-3, when power is applied to the circuit, flip-flops U_{5A} and U_{5B} usually start with the Q output of U_{5A} logic-low and the Q output of U_{5B} logic-high. The states of the

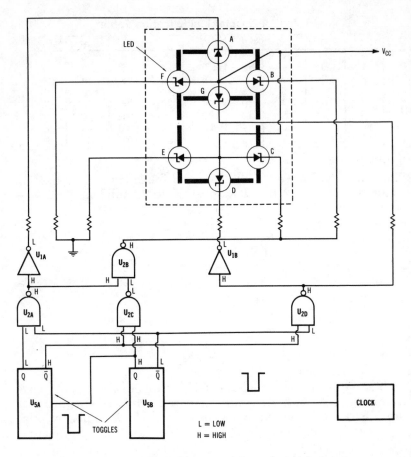

Fig. 12-3. Example of a binary-to-alphameric code converter.

logic devices are then as shown. When these states are present, it is apparent that the letter C will be displayed. Note that the segments E and F are not switched, but are continuously activated. A logic-low output from U_{1A} activates segment A, and a logic-low output from U_{1B} activates segment D. A logic-high output from NAND gate U_{2B} deactivates segments B and C, and a logic-high output from U_{2D} deactivates segment G.

When the first logic-low clock pulse is applied to U_{5B}, its Q output goes logic-low, and its $\overline{Q}$ output goes logic-high. When the U_{5B} Q output goes logic-low, flip-flop U_{5A} is clocked, driving its Q output logic-high and its $\overline{Q}$ output logic-low. In turn, segment A is deactivated, and an L is displayed. When the next logic-low clock pulse occurs, flip-flop U_{5B} changes state; its Q

260

output goes logic-high, and its $\overline{Q}$ output goes logic-low. However, flip-flop U_{5A} does not change state (it received a logic-high input). Note that flip-flop U_{5A} is counting in binary code; it changes state on every second clock pulse. Thus, the alphameric display sequence C, L, O, A is produced.

Gray-Code Converters

Input/output devices may require interfacing facilities that provide analog-to-digital (a/d) conversion. As an illustration, when a computer is used to predict the weather, one of the variables consists of the changing direction of a weather vane. This factor is processed by converting the changing direction into binary numbers for data entry into the computer. One widely used a/d converter of electromechanical design is shown in Fig. 12-4. The light areas in the disk represent a conducting material, whereas the dark areas represent an insulating material. Sensing of the rotational position is accomplished by means of three copper-carbon brushes, each of which contacts one concentric band. In this example, the converter employs eight segments. This three-bit encoder digitizes the shaft position into an equivalent three-bit binary number. The binary numbers are expressed in Gray code (Table 12-3). This code is widely utilized in a/d converters because it employs consecutive numbers that differ by a single bit. This format eliminates 180° ambiguity, and the greatest reading error is an ambiguity of one segment of rotation. Reading error may be reduced by using more segments. For example, an eight-bit converter provides a shaft resolution of 1.4°.

A converter for changing Gray code to binary code typically includes AND gates, OR gates, and inverters, as shown in Fig. 12-5. A three-bit converter is shown in this example. A con-

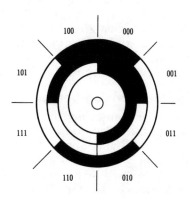

Fig. 12-4. A three-bit Gray-code a/d wheel.

Table 12-3. Gray Code

DECIMAL	BINARY	GRAY CODE
0	0000	0000
1	0001	0001
2	0010	0011
3	0011	0010
4	0100	0110
5	0101	0111
6	0110	0101
7	0111	0100
8	1000	1100
9	1001	1101
10	1010	1111
11	1011	1110
12	1100	1010
13	1101	1011
14	1110	1001

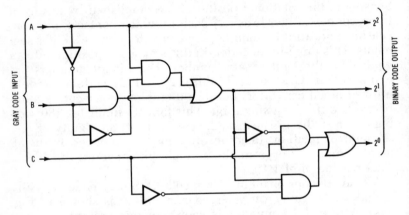

Fig. 12-5. Converter for changing Gray code into binary code.

verter for changing binary code into Gray code comprises a chain of XOR gates, as shown in Fig. 12-6. Thus, in the example shown in Fig. 12-6, a binary 101001 input is converted into a 111101 Gray output. A code converter is also called an *encoder*.

Digital-to-Analog Converter

A digital-to-analog converter (abbreviated d/a converter, or dac) is a configuration that changes digital words (such as those generated by an analog-to-digital converter) into a corresponding analog voltage. An example of a four-bit digital-to-analog converter is shown in Fig. 12-7. Four RS flip-flops are utilized as a storage register. The binary word that represents a given

262

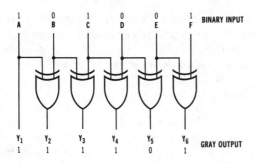

Fig. 12-6.   Converter for changing binary code into Gray code.

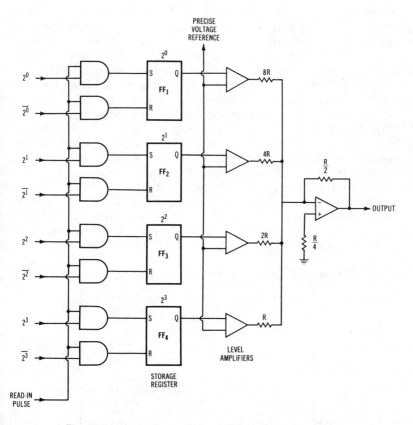

Fig. 12-7.   Logic diagram of a digital-to-analog converter.

263

voltage value is put into the storage register in parallel form; each flip-flop in the register is set to 0 or 1 when the read-in pulse is applied. As an illustration, if the input word is 1001, the 2^0 and the 2^3 flip-flops will be set to 1, and the 2^1 and 2^2 flip-flops will be set to 0 when the read-in pulse is applied. The output voltages from the register are passed through level amplifiers to ensure that precise 1 and 0 levels are applied to the op-amp circuit. The outputs from the level amplifiers are summed through a binary-weighted resistor network by the op amp. In turn, the op amp produces a precise output voltage corresponding to the voltage that is represented by the four-bit binary word applied at the input of the converter.

XS3 Code Conversion

The XS3 (excess-3) code is a nonweighted code. It is the same as binary coded decimal, except that the number 3 is added to each digit (Table 12-4). Note that in a weighted code each bit position has a certain value assigned to it, as in the pure binary and binary-coded-decimal formats. On the other hand, a nonweighted code, such as XS3, does not have assigned values for the bit positions. Encoding and decoding to and from the XS3 code are usually accomplished by adding 3 to each number when encoding and subtracting 3 from the number when decoding. An advantage of the XS3 code is that it is self-complementing; in other words, if all 0 and 1 bits in an XS3 number are complemented, the nines complement of the number is obtained. In some computer applications, this ability to obtain the nines complement reduces the hardware that is required to perform subtraction. Another advantage of the XS3 code is that all numbers have at least one 1 bit in them. This characteristic provides a distinction between zero and no information.

Converters for changing XS3 to bcd, XS3 to decimal, bcd to

Table 12-4. XS3 Code for Decimal and BCD Numbers

DECIMAL	BCD	XS3
0	0000	0011
1	0001	0100
2	0010	0101
3	0011	0110
4	0100	0111
5	0101	1000
6	0110	1001
7	0111	1010
8	1000	1011
9	1001	1100

XS3, or decimal to XS3 codes can be readily implemented with diode-logic matrix configurations. Note that the XS3 code is obtained by adding 3 to a decimal digit and then converting the sum to binary form.

A logic diagram for a four-bit XS3 adder is shown in Fig. 12-8. This arrangement processes one column of decimal digits. To add more than one column of decimal digits, XS3 adders are cascaded. The circuit in Fig. 12-8 receives two numbers in XS3 code and delivers their sum, also in XS3 code. In the example illustrated in Fig. 12-8, the two numbers are 1000 and 0101 (XS3 code for decimal 5 and 2, respectively), and their sum is 1010 (XS3 code for decimal 7).

MULTIPLEXERS

A multiplexer is used to rearrange the digital data on multiple input lines and to output the rearranged data on a single output line. Fig. 12-9 shows an example of an eight-channel multiplexer in which the data rearrangement, or multiplexing, is controllable by means of three input signals. In other words, the logic states present on any selected inputs, D_0 through D_7, are

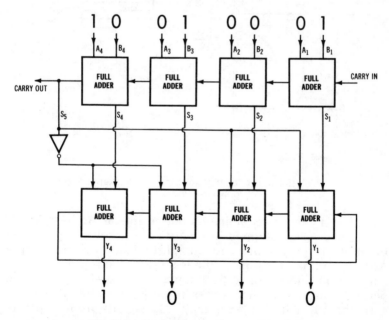

Fig. 12-8. Logic diagram for an XS3 adder.

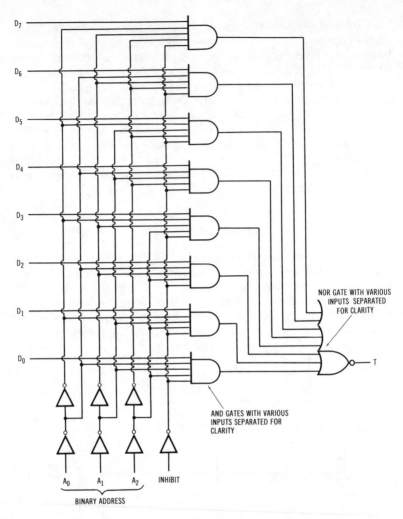

Fig. 12-9. Example of an eight-channel multiplexer. (*Courtesy Hewlett-Packard*)

placed in a chosen sequence on output line T. Address inputs A_0, A_1, and A_2 are stepped in binary code to select a particular input at the proper time. As an illustration, an address input $A_0=0$, $A_1=0$, $A_2=0$ will place the logic state that is present on D_0 onto output line T. An address input $A_0=1$, $A_1=0$, $A_2=0$ would place the logic state that is present on D_1 onto output line T.

In theory, a multiplexer may mix signals from multiple sources in various ways into a smaller number of outputs. Thus, the foregoing example of mixing eight data lines into a particular sequence on a single output line merely represents one widely used mode. Other multiplexers encountered in digital systems serve to multiplex four lines into one output line or 16 lines into one output line. There are digital word multiplexers that operate to multiplex three four-bit–wide parallel words into a single four-bit–wide output. With reference to Fig. 12-9, a fixed address may be applied so that the data from one input line is continuously channeled through to output line T. This operating mode of a multiplexer is termed a *data selector*.

An example of a dual four-input multiplexer IC package is shown in Fig. 12-10. Two address (select) inputs are provided, and complementary outputs are available. The select inputs are common to both of the multiplexers in this arrangement. Next, an eight-input multiplexer IC package is shown in Fig. 12-11. Three address (select) lines and an enable (inhibit) input are provided. When the enable input is logic-high, the multiplexer does not respond to data inputs; conversely, when the enable input is logic-low, the multiplexer responds as specified in the truth table. Complementary outputs are available.

DEMULTIPLEXERS

A *demultiplexer* has the opposite function from a multiplexer. That is, a demultiplexer receives data from a single source and distributes the data in accordance with a particular pattern into several output lines. For example, the demultiplexer shown in Fig. 12-12 has eight output lines and can accomplish the reverse of the data processing provided by the multiplexer in Fig. 12-11. In other words, serial data enters via input line T, the binary address (on A_0, A_1, and A_2) is sequentially stepped from 000 through 111, and the input data appears sequentially on output lines D_0 through D_7. Demultiplexer IC packages are available to distribute data from a single line into 2, 4, 16, and various other numbers of outputs.

Note that instead of being demultiplexed, data on the single input line (Fig. 12-12) can be gated through any one of the output lines continuously. This is accomplished simply by holding the address inputs on lines A_0, A_1, and A_2 constant. The address inputs can be changed when desired to route this data through another output line. The demultiplexer is called a *data distributor* in this mode of operation.

Observe also that a demultiplexer can be used as a decoder. In other words, if input line T (Fig. 12-12) is maintained logic-

267

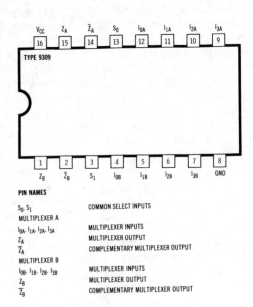

Vcc	Z_A	$\overline{Z}_A$	S_0	I_{0A}	I_{1A}	I_{2A}	I_{3A}
16	15	14	13	12	11	10	9

TYPE 9309

1	2	3	4	5	6	7	8
Z_B	$\overline{Z}_B$	S_1	I_{0B}	I_{1B}	I_{2B}	I_{3B}	GND

PIN NAMES

S_0, S_1	COMMON SELECT INPUTS
MULTIPLEXER A	
I_{0A}, I_{1A}, I_{2A}, I_{3A}	MULTIPLEXER INPUTS
Z_A	MULTIPLEXER OUTPUT
$\overline{Z}_A$	COMPLEMENTARY MULTIPLEXER OUTPUT
MULTIPLEXER B	
I_{0B}, I_{1B}, I_{2B}, I_{3B}	MULTIPLEXER INPUTS
Z_B	MULTIPLEXER OUTPUT
$\overline{Z}_B$	COMPLEMENTARY MULTIPLEXER OUTPUT

(A) Package pinout.

SELECT INPUTS		INPUTS				OUTPUTS	
S_0	S_1	I_{0A}	I_{1A}	I_{2A}	I_{3A}	Z_A	$\overline{Z}_A$
0	0	0	X	X	X	0	1
0	0	1	X	X	X	1	0
1	0	X	0	X	X	0	1
1	0	X	1	X	X	1	0
0	1	X	X	0	X	0	1
0	1	X	X	1	X	1	0
1	1	X	X	X	0	0	1
1	1	X	X	X	1	1	0
S_0	S_1	I_{0B}	I_{1B}	I_{2B}	I_{3B}	Z_B	$\overline{Z}_B$
0	0	0	X	X	X	0	1
0	0	1	X	X	X	1	0
1	0	X	0	X	X	0	1
1	0	X	1	X	X	1	0
0	1	X	X	0	X	0	1
0	1	X	X	1	X	1	0
1	1	X	X	X	0	0	1
1	1	X	X	X	1	1	0

X = EITHER HIGH OR LOW LOGIC LEVEL

(B) Truth table.

Fig. 12-10. A dual four-input multiplexer IC.

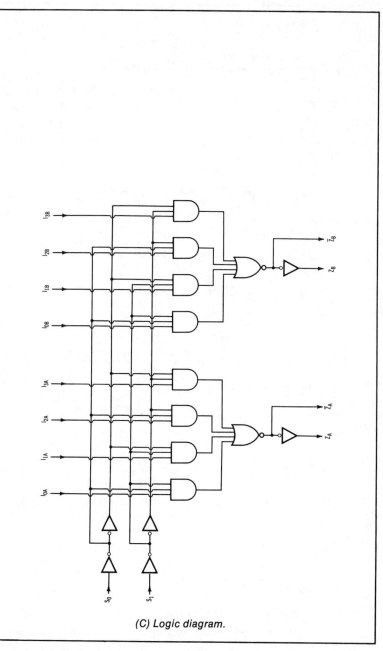

(C) Logic diagram.

(Courtesy Fairchild Camera and Instrument Corp.)

269

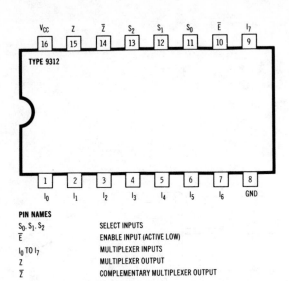

PIN NAMES

S_0, S_1, S_2	SELECT INPUTS
$\bar{E}$	ENABLE INPUT (ACTIVE LOW)
I_0 TO I_7	MULTIPLEXER INPUTS
Z	MULTIPLEXER OUTPUT
$\bar{Z}$	COMPLEMENTARY MULTIPLEXER OUTPUT

(A) Package pinout.

$\bar{E}$	S_2	S_1	S_0	I_0	I_1	I_2	I_3	I_4	I_5	I_6	I_7	$\bar{Z}$	Z
1	X	X	X	X	X	X	X	X	X	X	X	1	0
0	0	0	0	0	X	X	X	X	X	X	X	1	0
0	0	0	0	1	X	X	X	X	X	X	X	0	1
0	0	0	1	X	0	X	X	X	X	X	X	1	0
0	0	0	1	X	1	X	X	X	X	X	X	0	1
0	0	1	0	X	X	0	X	X	X	X	X	1	0
0	0	1	0	X	X	1	X	X	X	X	X	0	1
0	0	1	1	X	X	X	0	X	X	X	X	1	0
0	0	1	1	X	X	X	1	X	X	X	X	0	1
0	1	0	0	X	X	X	X	0	X	X	X	1	0
0	1	0	0	X	X	X	X	1	X	X	X	0	1
0	1	0	1	X	X	X	X	X	0	X	X	1	0
0	1	0	1	X	X	X	X	X	1	X	X	0	1
0	1	1	0	X	X	X	X	X	X	0	X	1	0
0	1	1	0	X	X	X	X	X	X	1	X	0	1
0	1	1	1	X	X	X	X	X	X	X	0	1	0
0	1	1	1	X	X	X	X	X	X	X	1	0	1

X = LEVEL DOES NOT AFFECT OUTPUT

(B) Truth table.

Fig. 12-11. An eight-input multiplexer IC.

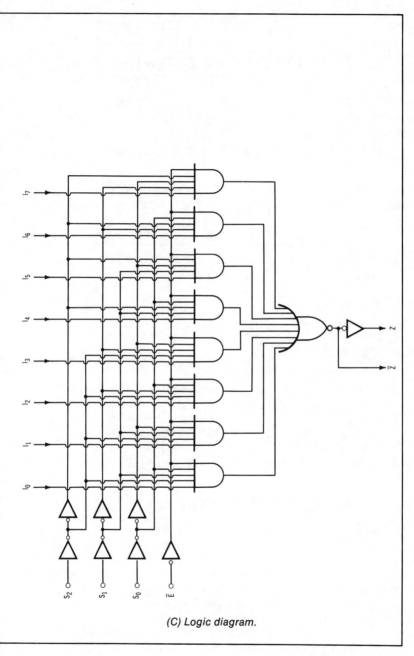

(C) Logic diagram.

(*Courtesy Fairchild Camera and Instrument Corp.*)

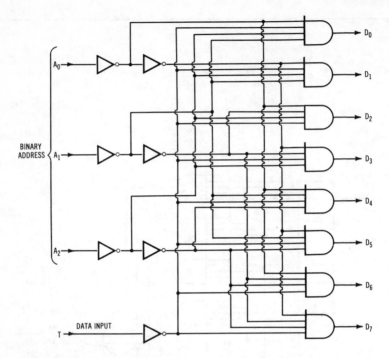

Fig. 12-12. An eight-channel demultiplexer. (*Courtesy Hewlett-Packard*)

low, the outputs will always represent the binary count corresponding to the logic levels on A_0, A_1, and A_2. For example, logic levels of $A_0=1$, $A_1=0$, and $A_2=1$ result in output line D_5 going high. (Line T becomes the enable function.)

It is evident that a demultiplexer with four address inputs and 16 outputs can be operated as a four-line–to–16-line decoder.

COMPARATORS

Digital comparators are widely used. A comparator functions to determine whether one of two binary numbers is larger than the other or if both numbers are equal. For example, we find comparators included in various adders and subtracters. Comparators are also utilized in digital control circuitry, wherein the control function that is generated depends upon the comparative values of two or more inputs. Binary addresses must also be compared for relative magnitude in some applications.

An example of a four-bit quad Exclusive-NOR comparator is shown in Fig. 12-13. This is a comparatively simple config-

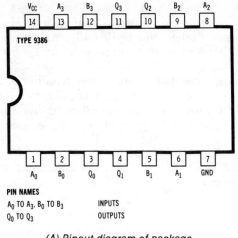

(A) Pinout diagram of package.

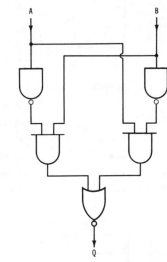

INPUTS		OUTPUT
A	B	Q
0	0	1
1	0	0
0	1	0
1	1	1

(B) Truth table (one section).   *(C) Logic diagram (one section).*

Fig. 12-13.   A quad XNOR comparator IC. (*Courtesy Fairchild Camera and Instrument Corp.*)

uration which determines only equality or inequality. As seen in the truth table, if both inputs are 1 or if both inputs are 0, the output is 1. On the other hand, if one input is 1 and the other input is 0, the output is 0.

An example of a more elaborate comparator is shown in Fig. 12-14. This is a five-bit configuration, and it determines whether

one digital word is equal to, greater than, or less than another
digital word. Thus, three outputs are provided. The comparator
responds as specified in the truth table. An active-low enable
input is included so that the comparator can be turned "on" or
"off."

A polarity (sign) comparator indicates whether two input voltages have the same polarity or opposite polarities. Two flip-flops
and two AND-OR gates are utilized in the configuration shown in
Fig. 12-15. If V_1 and and V_2 are both positive, gate A will have

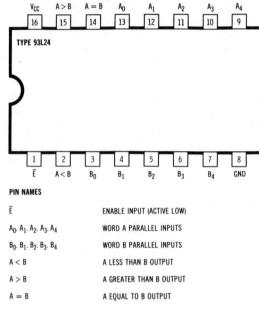

(A) Package pinout.

$\overline{E}$	A	B		A < B	A > B	A = B
1	X	X		0	0	0
0	WORD A = WORD B			0	0	1
0	WORD A > WORD B			0	1	0
0	WORD B > WORD A			1	0	0

X = "DON'T CARE"

(B) Truth table.

Fig. 12-14. A five-bit comparator IC.

274

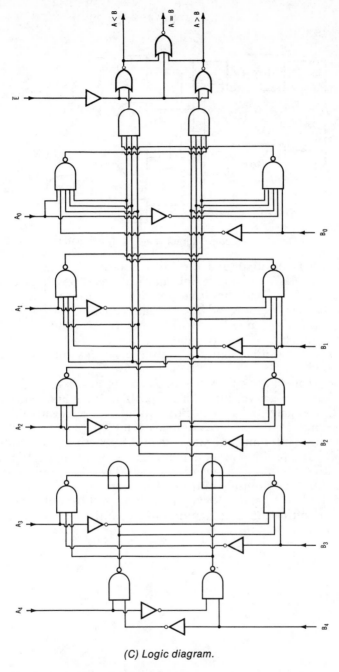

(C) Logic diagram.

(Courtesy Fairchild Camera and Instrument Corp.)

275

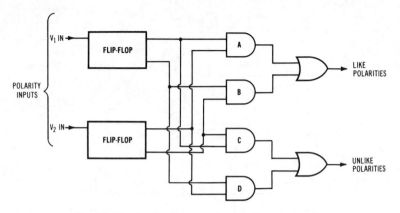

Fig. 12-15. Logic diagram of a polarity comparator.

both inputs positive, and the "like polarities" output will go logic-high. If V_1 and V_2 are both negative, gate B will have both inputs positive, and the "like polarities" output will go logic-high. If V_1 is positive and V_2 is negative, gate C will have both inputs positive, and the "unlike polarities" output will go logic-high. If V_1 is negative and V_2 is positive, gate D will have both inputs positive, and the "unlike polarities" output will go logic-high.

Note that when gate A has both inputs positive, gate B has both inputs negative, gate C has a positive input and a negative input, and gate D has a positive input and a negative input. Accordingly, only gate A produces a logic-high output. When gate A has both inputs negative, gate B has both inputs positive, gate C has a positive input and a negative input, and gate D has a positive input and a negative input. Therefore, only gate B produces a logic-high output. The same general principle applies when V_1 and V_2 have opposite polarities; only one gate can go logic-high for a particular combination of input polarities.

Chapter
13

Additional Counter Topics

In the first part of this chapter, additional counter configurations will be introduced. In the last part of the chapter, several useful troubleshooting pointers will be presented.

PROGRAMMABLE COUNTERS

A programmable counter is defined as any form of counter with a modulus or counting pattern which can be modified in some manner by a control signal. Common programming signals are related to presetting of the counter to a particular value (thereby changing the modulus of the counter), stopping the counter at a certain number, or causing the counter to reset itself and start counting over again. For example, Fig. 13-1 shows an asynchronous four-bit counter that can be preset to any desired count between 0 and 15. The desired number is input via NAND gates U_1 through U_4 by enabling the Load line. If the arrangement is to be operated as a modulus-7 counter, it would be preset to a count of 8, and it would then proceed to count to 15. Note that at the count of 15, the counter must again be preset to 8 (preset before the next clock pulse) if it is to repeat the same cycle. This requirement involves an additional load-control line.

VARIABLE-MODULO COUNTER

A widely used variable-modulo IC package is shown in Fig. 13-2. The arrangement consists of a three-stage counter and a single-stage binary counter. The three-stage section can be programmed (by means of the external connections listed in Chart 13-1) for operation as a modulo-5, 6, 7, or 8 counter. When this

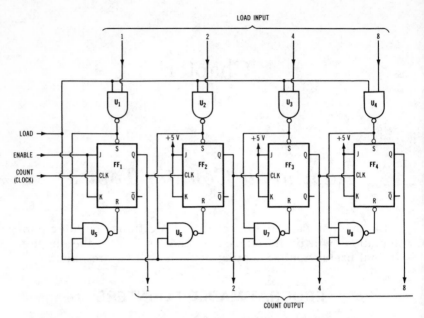

Fig. 13-1. Logic diagram of a programmable counter with preset inputs.
(*Courtesy Hewlett-Packard*)

counter is combined with the modulo-2 (binary) stage, modulos of 10, 12, 14, and 16 result. A four-stage binary counter can be formed by applying the input to the modulo-2 stage and then applying the $\overline{Q}$ output of this stage to the three-stage section. If an output with a 50% duty cycle is desired, the input may be applied to the three-stage section and then the $\overline{Q}_3$ output of this section applied to the modulo-2 section. (All clock inputs respond to the positive-going pulse transition.) There are asynchronous master set and reset imputs which act on all four stages. (Reset low drives all Q outputs low; set low drives all Q outputs high.) Note that a synchronous clear can be obtained by using the master set input since the next clock pulse after the master set pulse resets all four flip-flops. More than one counter can be cascaded by connecting the $\overline{Q}_3$ output to the input of the next (more significant) counter.

PRESELECTION OF FINAL COUNT

The synchronous four-bit counter shown in Fig. 13-3 will advance to a preselected count and hold there until the counter is reset. The modulus can be selected with the programming in-

278

puts via gates U_1 through U_4; however, it is the final count instead of the starting count that is determined. The principle that is utilized is to operate gates U_1 through U_4 to make a digital comparison; when the output of the counter matches the programming inputs, the clock is disabled and the circuit stops counting. At this point, the Count Complete signal can be used to reset the counter, or the contents of the counter can be saved.

COUNTER VARIATIONS

The counter configurations discussed so far employ the 8421 binary code. These counters are the most common; they are the simplest to build (require the fewest gates to interconnect the flip-flops), and they use the flip-flops most efficiently (can count to the highest number compared to the number of flip-flops). However, any of the asynchronous and synchronous counters can be configured to count in codes other than 8421 binary.

The nucleus of any counter is a series of flip-flops. An 8421 counter differs from any other type of counter only in the circuitry that is used to interconnect the flip-flops. Fig. 13-4 shows an example of interconnecting circuitry that is different from that used in an 8421 binary counter. This four-bit counter counts in 2421 binary code, and it operates as an excess-3 code counter if pin 1 of gate U_3 is connected to Q of FF_4 instead of $\overline{Q}$.

Counters whose moduli are not powers of 2 require a characteristic type of gating. Examples are shown in Figs. 13-5 through 13-8.

DIGITAL ELECTRONICS IN HOUSEHOLD APPLIANCES

A timer is the "heart" of many household appliances. For example, timers are used to turn on lights, to turn off lights, to control ovens, to cycle mechanical operations, to control hydraulic systems, and so on. A diagram for a basic timer arrangement is shown in Fig. 13-9. The IC package is detailed in Fig. 13-10. Observe that the timer can be started with push button SW_3 and reset with push button SW_2. The length of time for which the timer is set is determined by adjustment of variable resistor R_1. Two outlets are provided on the timer; one outlet is normally on and will turn off when the timer reaches the end of the set interval. The other outlet is normally off and will turn on when the timer reaches the end of the set interval.

To start the timing cycle again, the reset button is pressed momentarily. If external terminals are connected across SW₃ and SW₂, the timer can be turned on or reset by remote control, using either a momentary push-button switch or a relay activated by a photocell, thermostat, humidity sensor, smoke detector, etc. Digital pulses from other circuits, such as a personal computer, can be used to trigger the time of starting or of resetting. A capacitor must be used to couple external pulses to the trigger and reset terminals.

With reference to Fig. 13-10, the IC functions as a highly

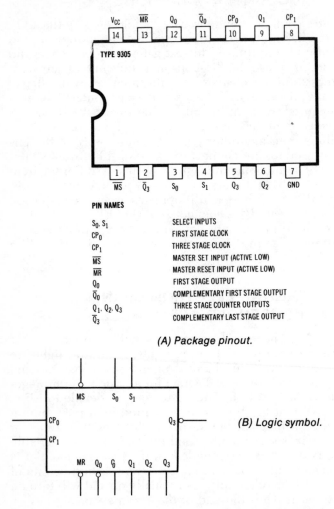

(A) Package pinout.

(B) Logic symbol.

Fig. 13-2.   A variable-modulo counter IC.

280

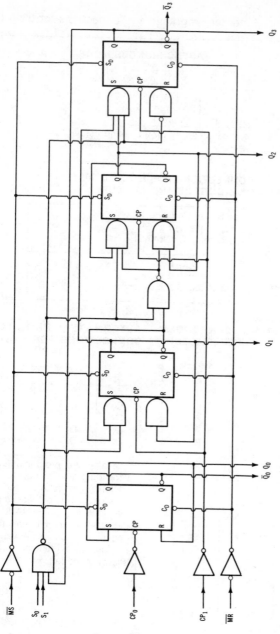

(C) Logic diagram.

(Courtesy Fairchild Camera and Instrument Corp.)

281

Chart 13-1. External Programming Connections for 9305 Counter

ASYNCHRONOUS MODE

$\overline{MS}$	$\overline{MR}$	Q_0	$\overline{Q}_0$	Q_1	Q_2	Q_3	$\overline{Q}_3$
0	1	1	0	1	1	1	0
1	0	0	1	0	0	0	1
1	1	Count*					

*As determined by programming connections.

CONNECTIONS FOR MODULO 10, 12, 14, 16
BINARY COUNTERS AND 50% DUTY CYCLE DIVIDERS

For Binary Counting
$\overline{Q}_0$ connected to CP_1
Incoming clock to CP_0

For 50% Duty Cycle Output
$\overline{Q}_3$ connected to CP_0
Incoming Clock to CP_1

PROGRAMMING CONNECTIONS FOR LAST THREE STAGES

S_0	S_1	Modulo
NC	NC	5
Q_1	NC	6
NC	Q_1	6
Q_2	NC	7
NC	Q_2	7
Q_1	Q_2	8
Q_2	Q_1	8

NC = Not Connected

Modulo	Output	Available Output Fan-Out
5	Q_3	14/8
6	Q_1	14/7
7	Q_2	14/7
8	Q_1	15/7
8	Q_2	15/7

ALTERNATE** PROGRAMMING CONNECTIONS FOR LAST THREE STAGES

S_0	S_1	Modulo
Q_3	Q_3	5
Q_1	Q_1	6
Q_2	Q_2	7
Q_1	Q_2	8
Q_2	Q_1	8

**The alternate programming connections program the counter and conveniently terminate unused select inputs (NC). Since these inputs form the inputs to a single NAND gate (see Fig. 13-2C), their connection to the counter outputs for the various count modulos provides the output drive shown at the left.

Courtesy Fairchild Camera and Instrument Corp.

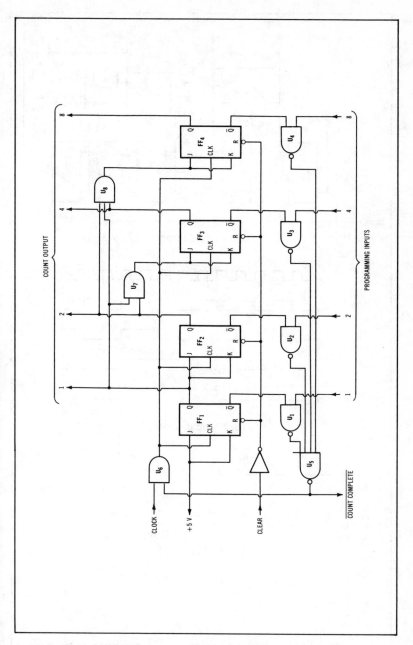

Fig. 13-3. Programmable counter with preselection of final count.
(*Courtesy Hewlett-Packard*)

283

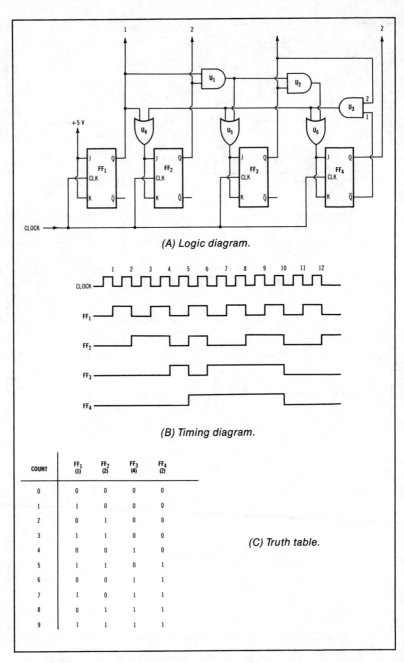

(A) Logic diagram.

(B) Timing diagram.

COUNT	FF₁ (1)	FF₂ (2)	FF₃ (4)	FF₄ (2)	
0	0	0	0	0	
1	1	0	0	0	
2	0	1	0	0	
3	1	1	0	0	
4	0	0	1	0	(C) Truth table.
5	1	1	0	1	
6	0	0	1	1	
7	1	0	1	1	
8	0	1	1	1	
9	1	1	1	1	

Fig. 13-4. A synchronous 2421 counter. (*Courtesy Hewlett-Packard*)

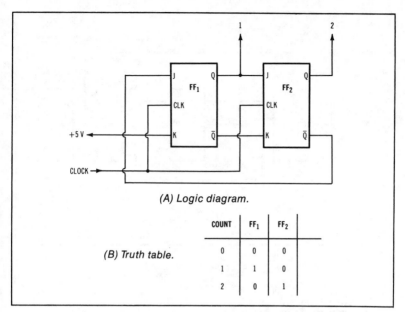

(A) Logic diagram.

(B) Truth table.

COUNT	FF$_1$	FF$_2$
0	0	0
1	1	0
2	0	1

Fig. 13-5. A modulo-3 counter. (*Courtesy Hewlett-Packard*)

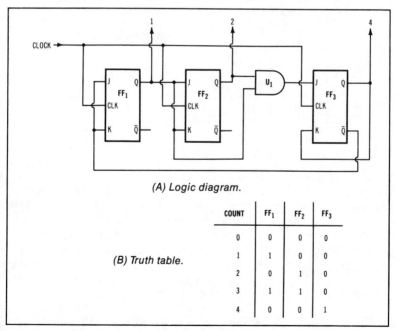

(A) Logic diagram.

(B) Truth table.

COUNT	FF$_1$	FF$_2$	FF$_3$
0	0	0	0
1	1	0	0
2	0	1	0
3	1	1	0
4	0	0	1

Fig. 13-6. A modulo-5 counter. (*Courtesy Hewlett-Packard*)

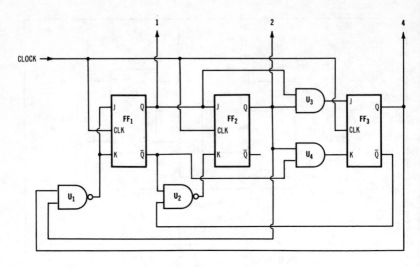

(A) Logic diagram.

COUNT	FF$_1$	FF$_2$	FF$_3$
0	0	0	0
1	1	0	0
2	0	1	0
3	1	1	0
4	0	0	1
5	1	0	1
6	0	1	1

(B) Truth table.

Fig. 13-7.   A modulo-7 counter.

stable controller capable of providing accurate time delay. The time interval is precisely determined by one external resistor and one external capacitor. This configuration is trailing-edge triggered. It consists of two comparators, a flip-flop, and an output stage. Each comparator is a high-gain operational amplifier that provides precise comparison of two voltages. If one input voltage rises to the threshold value established by the other input voltage, the output from the comparator suddenly changes polarity and triggers the flip-flop.

Fig. 13-11 shows the basic external connections for timer operation of the IC. The timing process is as follows: The external capacitor (C) is initially held discharged by a transistor inside

286

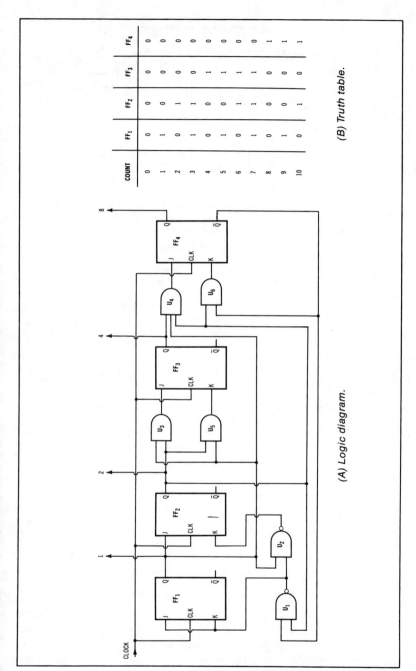

(A) Logic diagram.

(B) Truth table.

Fig. 13-8. A modulo-11 counter.

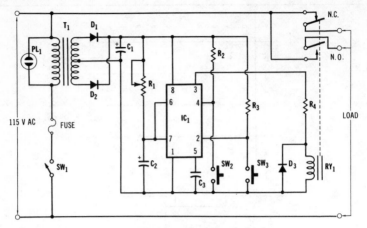

Fig. 13-9. Diagram of a household timer.

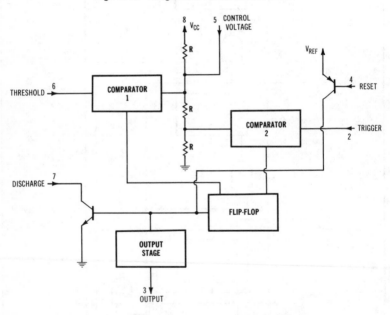

Fig. 13-10. Logic diagram of IC for Fig. 13-9.

the IC (Fig. 13-10). Upon application of a negative trigger pulse
to pin 2, the flip-flop is set; this removes the short circuit across
external capacitor C and drives the output high. Now, the voltage across the capacitor increases exponentially in accordance
with its time constant (T = RC, or time in seconds equals ohms

288

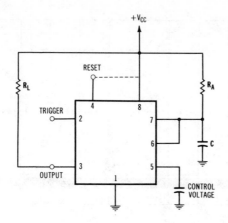

Fig. 13-11.   Basic external connections for timer IC.

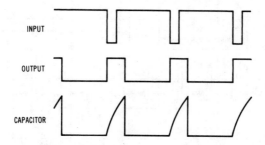

Fig. 13-12.   Typical timer waveforms.

times farads). When the voltage across the capacitor equals 2/3 V_{CC}, comparator 1 resets the flip-flop, which in turn discharges the capacitor suddenly and drives the output to its low state. The waveforms that are normally generated are shown in Fig. 13-12.

This circuit triggers on a negative-going input signal when its level reaches 1/3 V_{CC}. Once triggered, the circuit will remain in this state until the set time has elapsed, even if it is triggered again during this interval. Since the charge rate and the threshold level of the comparator are both directly proportional to the supply voltage, the timing interval is independent of the supply voltage. Application of a negative pulse simultaneously to the reset terminal (pin 4) and to the trigger terminal (pin 2) during the timing cycle discharges the external capacitor and causes the cycle to start over again. The timing cycle will now

289

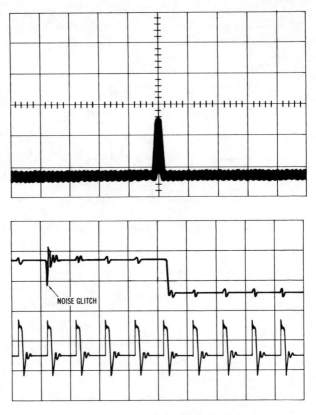

Fig. 13-13. Typical noise-glitch displays.

start on the positive edge of the reset pulse. During the time that the reset pulse is applied, the output is driven to its low state. Note that when the reset function is not in use, it is good practice to connect pin 4 to V_{CC} in order to avoid the possibility of false triggering. Troubleshooting of the timer circuit should be done by waveform analysis with a triggered-sweep oscilloscope.

NOISE GLITCHES IN
DIGITAL WAVEFORMS

Noise glitches (Fig. 13-13) usually are of longer duration and are more visible in a digital waveform than are race glitches. Noise voltages may have an internal source or an external source. For example, although an electron travels very rapidly, in a large digital system there are propagation delays from input to output which are easily measurable with a triggered-sweep oscilloscope. Propagation delay causes data changes to occur at slightly different times on the data lines, resulting in internal noise. One line may pick up noise from another line. Noise voltages may have an external source in industrial applications; marginal power supplies can also develop glitches. In most situations, an oscilloscope can be used to advantage as a glitch-tracer for tracking down a noise glitch to its source.

PRECAUTIONS IN
DIGITAL TROUBLESHOOTING

Experienced digital troubleshooters generally agree that no more than two technicians should work on a malfunctioning system at the same time. Having more than two workers can lead to unjustified assumptions, misunderstanding, and confusion. In many trouble situations, there will *appear* to be more than one fault or malfunction. However, digital trouble symptoms are often interrelated. For example, low line voltage aggravated by line-voltage variation can make it appear that several faults are present in a digital system. It is good practice to correct the "easy" faults first; it may then happen that the more obscure trouble symptoms disappear. Thus, if malfunction *could* be caused by low line voltage or fluctuating line voltage, this "easy" fault condition should be checked out before you proceed further.

It is helpful to check the operation of a digital system on all of its functions and to list all recognizable trouble symptoms before "getting into the circuitry." When the trouble symptoms are "sized up," it may be found that there is a common factor linking most or all of the symptoms together. For example, in Chart 13-2 it can be seen that low sensitivity, noise output but no reception, failure of the scanning logic with the squelch inoperative, and dark LED with no sound output have a weak battery as a common cause. Some digital systems include peripherals; those that are not involved in the malfunction may be unplugged. The neophyte tends to make unjustified assumptions, such as "the defect must be in. . ." or that "the waveform must be normal." In other words, evidence should be obtained for the existence of a particular signal within the trouble area, and oscilloscope observations should be used to verify that the signal is normal.

291

Chart 13-2. Troubleshooting Chart for a Four-Channel Scanner Monitor Radio Receiver

Symptom	Possible Cause
Light-emitting diode (LED) does not light and no sound is audible Power Switch: On Volume: Maximum Channel Switch: On	A. Defective/low/weak battery B. Defective on-off switch on volume control C. Power circuit disconnected or short circuit
LED lights but no sound Channel Switch: On Squelch: Off (Minimum) Volume: Maximum	A. Defective earphone jack B. Defective speaker C. Faulty transistor or related circuit component
Sound but LED does not light Channel Switch: On Squelch: Off (Minimum) Volume: Maximum	A. Faulty LED pc board assembly +B supply B. Defective LED C. Faulty channel switch D. Defective IC
No sound through earphone	A. Defective earphone jack B. Defective earphone
Does not scan and squelch does not operate	A. Faulty noise amplifier or faulty noise detector circuit component B. Defective rf or if amplifier C. Defective/low/weak battery
Does not scan but squelch operates normally	A. Defective auto/manual switch B. Defective transistors C. Defective IC D. One of circuit component parts faulty
Manual selector does not operate	A. Defective auto/manual selector switch B. Defective transistor C. Defective IC D. One of circuit component parts faulty
Skipper circuit does not operate	A. Faulty transistor or circuit component part
Only scans channels 1 and 2 or channels 3 and 4	A. Defective IC
LED of one or two channels does not light when channel switch is on	A. Defective IC B. Defective channel switch C. Defective LED

Chart 13-2. Troubleshooting Chart for a Four-Channel Scanner Monitor Radio Receiver (cont)

Noise can be heard but does not receive	A. Wrong crystal channel or faulty crystal B. Defective rf amplifier circuit C. Defective if amplifier circuit D. Defective 1st or 2nd local oscillator circuit E. Weak battery
Mixed reception	A. Crystal not set to proper receiving frequency
Low sensitivity	A. Weak battery B. Crystal frequency not covering the correct channel C. Defective rf and if amplifiers D. Retune front end higher or lower to obtain optimum sensitivity for the frequency change desired.

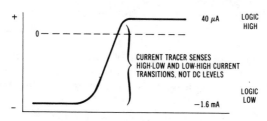

(A) Levels in standard TTL (fan-out of one).

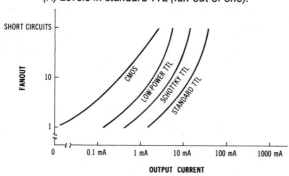

(B) Comparison of common logic families.

Fig. 13-14. Current levels in normal and shorted digital circuits.
(Courtesy Hewlett-Packard)

TROUBLESHOOTING CONSIDERATIONS

A logic current tracer is an ac-responding device. A current probe detects and displays current pulses or transitions and then stretches the pulses and displays them by means of its indicator lamp. When a TTL output goes from logic low to logic high, for example, the total current change is about 1.6 mA (Fig. 13-14A). Note that a current tracer is not voltage-sensitive, and that it will respond only to current changes. As shown in Fig. 13-14B, CMOS, low-power TTL, and Schottky TTL circuits operate at somewhat lower current levels than standard TTL circuits. However, in all cases, the short-circuit current level is typically 10 times the normal operating current level. The current tracer of Fig. 13-15 can be set for a sensitivity from 1 mA to 1 A.

Fig. 13-16 shows LED indications for various sensitivity settings and the technique for estimating current levels. In typical digital troubleshooting situations, a current tracer is the most important test instrument that will be used to pinpoint a fault on a node. Usually, the current tracer is used to locate a fault by following the current after voltage or logic-state sensing tests have been made to narrow the trouble area down to a node, pc board conductor, or bus line (main digital information path). Then, since the area under test is probably stuck at a fixed voltage level, only current-tracing tests will serve to indicate an activity path that can be followed. In other words, a node "stuck" in one state (high or low) may be trying very hard to change

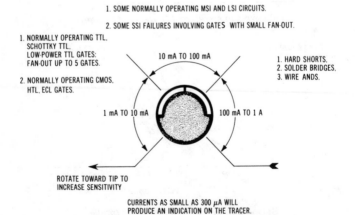

Fig. 13-15. Interpretations of current-tracer sensitivity settings.
(*Courtesy Hewlett-Packard*)

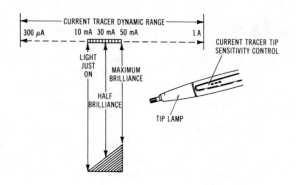

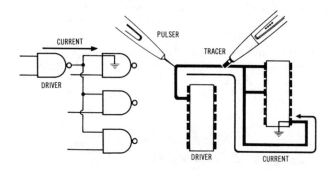

CURRENT NEEDED FOR BARELY LIGHTED LAMP	CURRENT WHEN LAMP SET FOR HALF-BRILLIANCE	CURRENT THAT WILL PRODUCE A FULLY BRIGHT LAMP
300 μA	1 mA	≥ 3 mA
1 mA	3 mA	≥ 5 mA
3 mA	5 mA	≥ 10 mA
5 mA	10 mA	≥ 30 mA
10 mA	30 mA	≥ 50 mA
30 mA	50 mA	≥ 100 mA
50 mA	100 mA	≥ 500 mA
300 mA	500 mA	≥ 1 mA

WHEN A 10 mA CURRENT TRANSITION OCCURS AND SENSITIVITY IS SET FOR HALF-BRILLIANCE OF THE TIP LAMP, THE DYNAMIC RANGE OF THE CURRENT TRACER IS AS FOLLOWS:

DIM LAMP ....5 mA
HALF-BRIGHT LAMP................10 mA
FULLY LIGHTED LAMP ....≥ 30 mA

THE TABLE ABOVE SHOWS SEVERAL EXAMPLES OF DYNAMIC RANGE FOR THE FULL RANGE OF POSSIBLE SENSITIVITY CONTROL SETTINGS AVAILABLE.

Fig. 13-16.   Interpretation of current-tracer indications.
(*Courtesy Hewlett-Packard*)

Fig. 13-17.   Troubleshooting a "stuck-at" node. (*Courtesy Hewlett-Packard*)

state, and in turn the node will be carrying a large amount of current.

In Fig. 13-17, the node is "stuck low," and the technician needs to determine whether the driver is dead or a shorted input is clamping the node to ground. Accordingly, a logic probe and pulser are applied to test the logic state of the node and whether or not it can be changed. For example, shorts to V_{CC} or to ground cannot be overridden by pulsing. However, injection of pulses at the node enables the technician to trace the current directly to the fault point. Current-tracing tests are facilitated by adjusting the tracer sensitivity to a point such that the indicator light is just visible with the logic pulser set to its 100-Hz mode. As the current path is being traced, the indicator light will suddenly become dark; this indicates that the current tracer has moved directly over the fault point.

CURRENT-TRACING TECHNIQUES

Printed-circuit conductors that vary greatly in width (Fig. 13-18) cause flux-density changes under the tip of a current tracer. This flux-density variation can be of practical importance in tracing supply-to-ground shorts; the sensitivity of the current tracer may need to be changed slightly for optimum indication. Setting an appropriate reference on a node identified as faulty is a fundamental adjustment in the current-tracing procedure. Note that the reference setting for one node has little, if any, relevance for other nodes, due to fan-out (branched loads) or variability in circuit interconnections. Also, the sensitivity control on the current tracer allows the technician to "see" currents as small as 300 μA, but there is virtually no upper limit. If the sensitivity control is set so that 10 mA barely lights the tracer display, 30 mA will produce half brilliance, and full brilliance is reached at 50 mA or slightly more. Currents greater than 50 mA also produce full brilliance.

If a current path seems to disappear as the current tracer is moved along a conductor, observe whether the conductor becomes wider, resulting in a spreading out of the current and a lessening of the field intensity under the tracer tip. In such a

Fig. 13-18.   Effect of conductor width on current density.
(*Courtesy Hewlett-Packard*)

case, the sensitivity of the current tracer should be increased. Sometimes it may be observed that the conductor proceeds down a plated-through hole in the pc board. This means that the tip of the tracer is then farther from the current; therefore, the sensitivity of the tracer should be increased. In other cases, it may be observed that the current path "branches" and proceeds to several different places via several different paths. A branch point corresponds to a sudden reduction of field intensity. Accordingly, the tracer sensitivity should be increased to obtain an adequate indication.

Note that a logic probe cannot be used in this type of trouble situation because a short creates a very low circuit impedance. When the impedance is very low, a large current change corresponds to a very small voltage change. Such small voltage changes are beyond the response capability of a logic probe; however, the large current changes are easily checked with a current tracer.

Chapter
14

Addition, Multiplication, Division, and Video Games

This chapter will show some of the ways for implementing the operations of addition, multiplication, and division. Also, video games are discussed, and additional troubleshooting hints are given.

LOOK-AHEAD–CARRY ADDER

A look-ahead–carry adder is a rapid parallel adder that does not have to wait for the carries to ripple through from one internal full-adder stage to the next. A look-ahead configuration uses auxiliary logic to anticipate all carry signals before they are actually generated. This auxiliary logic provides correct carry inputs simultaneously to all stages of the adder, before they are developed separately by each stage; it also anticipates what the carry output of the last full adder will be. For this reason, look-ahead–carry adders are significantly faster than parallel adders with ripple carry. An example of a four-bit look-ahead–carry adder configuration with carry anticipation logic is shown in Fig. 14-1. This arrangement is called a *zero-level look-ahead–carry* adder.

An example of a look-ahead–carry generator IC is shown in Fig. 14-2. It is generally used with a four-bit arithmetic logic unit (ALU) to provide high-speed look-ahead over word lengths of more than four bits. The look-ahead–carry generator accepts up to four pairs of active-low carry-propagate ($\overline{P}_0$, $\overline{P}_1$, $\overline{P}_2$, $\overline{P}_3$) and carry-generate ($\overline{G}_0$, $\overline{G}_1$, $\overline{G}_2$, $\overline{G}_3$) signals and an active-high carry input (C_n); it provides active-high carries (C_{n+x}, C_{n+y}, C_{n+z}) across four groups of binary adders.

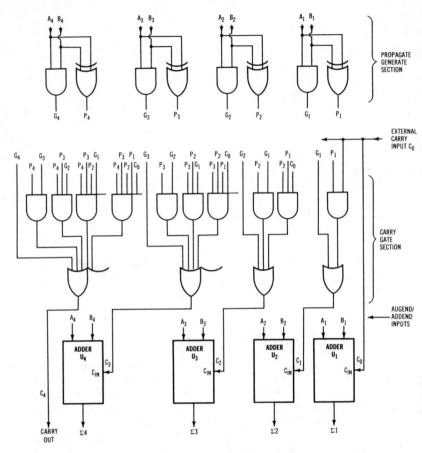

Fig. 14-1. Configuration of a look-ahead–carry adder. (*Courtesy Hewlett-Packard*)

BINARY-CODED–DECIMAL ADDERS

In digital calculators and voltmeters, binary-coded-decimal (bcd) numbers are often used. A bcd adder is not greatly different from the binary adders that were discussed previously; however, the data is processed in bcd format, as shown in Fig. 14-3. Observe the distinction between this bcd format and the straight binary format:

In bcd format, 589 = 0101 1000 1001
In straight binary format, 589 = 1001001101

Decimal addition often requires the carrying of a 1. For example, if 243 is added to 386, a 1 must be carried to arrive at the sum of 629. However, when this addition is processed in bcd format, the addition of each of the groups separately yields the following result:

386	0011	1000	0110
243	0010	0100	0011
629	0101	1100	1001
	5	12	9

Of course, 5 12 9 is not the same notation as 629. Correct processing requires that the 2 in 12 be written in the tens position with the 1 carried to the hundreds position, thereby changing the 5 into a 6. Therefore, when numbers in bcd format are added by logic circuitry, special auxiliary logic is necessary to detect a carry and to correct the bcd digit that generated the carry. In the foregoing example, beyond detection of the carry and its propagation to the hundreds group, the tens group must be corrected in order to output 2 in binary instead of 12 (it must be changed from 1100 to 0010). A bcd adder circuit that can generate the carry and correct the adder output of one bcd digit (nibble, or four bits) is shown in Fig. 14-4.

Note that even though a four-bit binary full-adder circuit can add two numbers whose sum is as large as 15 without generating a carry, both a carry and the accompanying correction are necessary whenever the sum is 10 or greater (corresponding to addition of decimal numbers). In the foregoing example, only the tens group generated a carry and necessitated a correction; in the general case, however, any one of the digit groups may exceed 9 during an addition procedure. For example, to add two three-digit bcd numbers together, three parallel adder stages are needed, each provided with the ability to generate a carry and to correct its output when necessary.

MULTIPLICATION

As noted previously, multiplication is repeated addition; thus, $5 \times 3 = 5 + 5 + 5$. The four rules, or algorithms, of binary multiplication are:

$$0 \times 0 = 0$$
$$0 \times 1 = 0$$
$$1 \times 0 = 0$$
$$1 \times 1 = 1$$

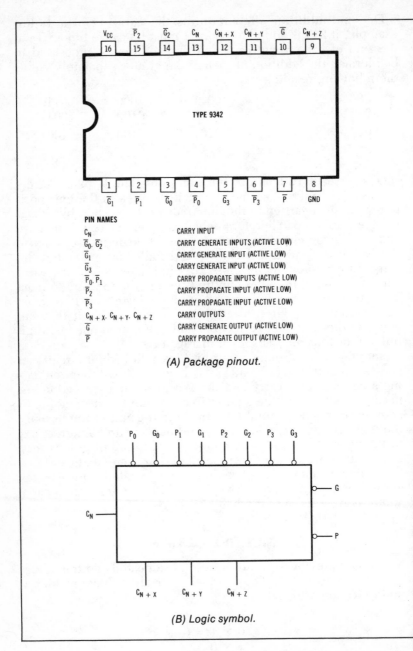

PIN NAMES

C_N	CARRY INPUT
$\overline{G}_0$, $\overline{G}_2$	CARRY GENERATE INPUTS (ACTIVE LOW)
$\overline{G}_1$	CARRY GENERATE INPUT (ACTIVE LOW)
$\overline{G}_3$	CARRY GENERATE INPUT (ACTIVE LOW)
$\overline{P}_0$, $\overline{P}_1$	CARRY PROPAGATE INPUTS (ACTIVE LOW)
$\overline{P}_2$	CARRY PROPAGATE INPUT (ACTIVE LOW)
$\overline{P}_3$	CARRY PROPAGATE INPUT (ACTIVE LOW)
C_{N+X}, C_{N+Y}, C_{N+Z}	CARRY OUTPUTS
$\overline{G}$	CARRY GENERATE OUTPUT (ACTIVE LOW)
$\overline{P}$	CARRY PROPAGATE OUTPUT (ACTIVE LOW)

(A) Package pinout.

(B) Logic symbol.

Fig. 14-2.   A look-ahead–carry generator IC.

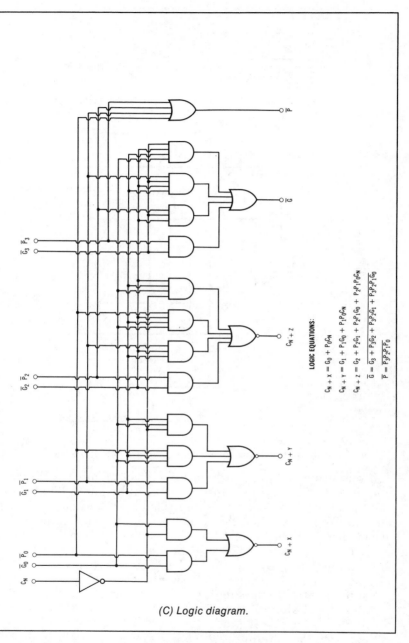

LOGIC EQUATIONS:

$$C_N + x = G_0 + P_0 C_N$$

$$C_N + y = G_1 + P_1 G_0 + P_1 P_0 C_N$$

$$C_N + z = G_2 + P_2 G_1 + P_2 P_1 G_0 + P_2 P_1 P_0 C_N$$

$$\overline{G} = \overline{G_3 + P_3 G_2 + P_3 P_2 G_1 + P_3 P_2 P_1 G_0}$$

$$\overline{P} = \overline{P_3 P_2 P_1 P_0}$$

(C) Logic diagram.

(Courtesy Fairchild Camera and Instrument Corp.)

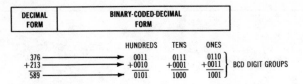

DECIMAL FORM	BINARY-CODED-DECIMAL FORM		
	HUNDREDS	TENS	ONES
376 ⟶	0011	0111	0110
+213 ⟶	+0010	+0001	+0011
589 ⟶	0101	1000	1001

BCD DIGIT GROUPS

Fig. 14-3.   Addition of two numbers in decimal and bcd form.

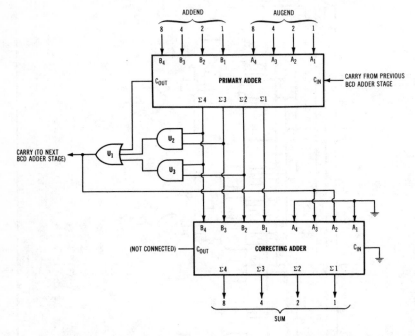

Fig. 14-4.   A typical bcd adder stage. (*Courtesy Hewlett-Packard*)

It was also noted previously that if a binary number is shifted one place to the left in a register, it is thereby multiplied by 2. Multiplication is ordinarily accomplished by means of an adder and shift registers called the A register, the B register, and the accumulator, as shown in Fig. 14-5. Since all binary numbers consist of 1s and 0s, each step in multiplication consists of either multiplying by 1 or multiplying by 0. When the multiplicand is multiplied by 1, it is simply repeated and shifted; when the multiplicand is multiplied by 0, and addition of 0000 is entered and shifted. Note in Fig. 14-5 how the adder forms the sums of the partial products until the final product is obtained. Note that shifting each partial sum to the right has the same effect as shifting the next partial product to the left.

304

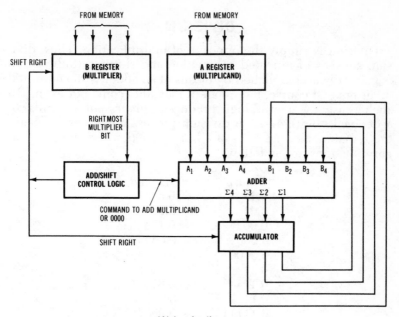

(A) Logic diagram.

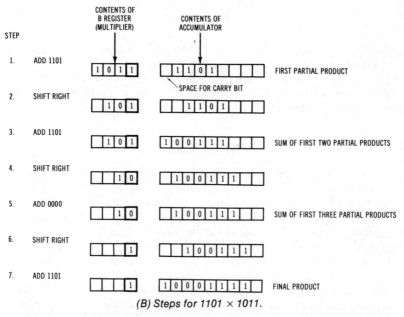

(B) Steps for 1101 × 1011.

Fig. 14-5. Example of a binary multiplier. (*Courtesy Hewlett-Packard*)

DIVISION

Division is simply the opposite of multiplication. Thus, division consists of repeated subtraction and shifting. (Subtraction is generally accomplished by means of an adder with controlled inverters.) Of course, the quotient may not "come out even." In such a case, the control logic produces a *binary point* (similar to a decimal point), and the division process continues to obtain the remainder in the form of a binary fraction. Examples of binary fractions are as follows:

Binary	*Decimal*
0.1	0.5
0.01	0.25
0.001	0.125
0.0001	0.0625
0.00001	0.03125
0.11	0.75
0.111	0.875
0.1111	0.9375
0.11111	0.96875
0.101	0.625
0.1001	0.5625
0.10101	0.65625

In other words, the first place to the right of the binary point has a value of ½; the second place to the right has a value of ¼; the third place to the right has a value of ⅛; and so on, with each denominator being the next higher power of 2. This simple relationship makes it easy to read binary fractions.

VIDEO GAMES

Video games employ digital logic to simulate sports events, outer-space warfare, and various "action" situations. Digital pulses are added to the horizontal-sync, vertical-sync, and blanking pulses of a tv signal to simulate lines, players, balls, vehicles, missiles, space ships, and so on. The composite video signal is modulated on a channel 3 or channel 4 carrier; a few video games also provide a sound signal. A typical video game utilizes four LSI ICs. Some of the output signals produce continuous lines or bars on the tv screen; other signals simulate a "ball" and "players," for example, and these can be moved about the screen during the game under joystick control. In addition, video data for scoring may be provided; digital readout is occasionally provided on the screen. Fig. 14-6 is a block diagram of one type of video game.

A joystick is a control handle with X and Y axes of movement; a pair of potentiometers are usually attached to a gimbal. Since a joystick is an analog device, analog-to-digital converters are required to interface the operator control to the digital-logic circuitry. A "spot" or "ball" is generally produced by applying a vertical line (or bar) and a horizontal line (or bar) to an AND gate; coincidence of the lines results in an output pulse from the gate, which in turn produces a visible spot on the screen. Note that a fixed line may not be continuous; it may have a discontinuity to represent a "goal," for example. Players are typically simulated by narrow rectangles. Each player in the game usually has three joysticks available; a horizontal joystick controls motion of the player spot from left to right; a vertical joystick controls motion of the player spot up and down; a ball joystick adds vertical motion to the ball spot. "Rebound" logic is also provided for "ball" motion; a pair of voltage comparators activate a switching transistor at appropriate limits of "ball" travel and reverse its direction of motion.

With reference to Fig. 14-7, each sync generator in a typical arrangement employs a pair of Schmitt triggers with an AND gate and an inverter. Each spot generator (Fig. 14-8) produces a video pulse that has a position on the screen which is determined by the value of a position control voltage (joystick control). A sync pulse is inputted with the position control voltage into a timing capacitor which actuates a pair of voltage comparators. Comparator outputs are applied to an AND gate, with the result that a video pulse is produced which occupies a position between successive sync pulses as determined by the value of the position control voltage.

CURRENT-TRACING OPERATING FACTORS

Adjustment of the sensitivity control may be used as a method of alleviating "crosstalk" interference when troubleshooting with a current tracer. The technician should also keep in mind that the tracer tip is directional. In other words, current paths oriented 90° with respect to the pickup coil tend to null out. Therefore, proper tip orientation (alignment) helps to eliminate crosstalk from conductors on different layers of a board or conductors at different angles. Tracer orientation is illustrated in Fig. 14-9. In areas where a pc board has many conductors side

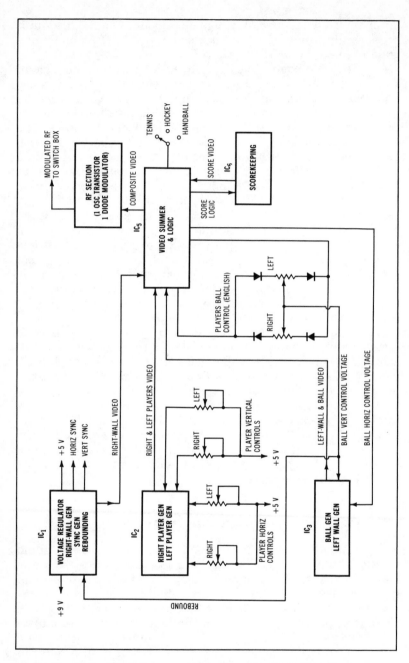

Fig. 14-6. Block diagram of Magnavox Odyssey 200 video game.

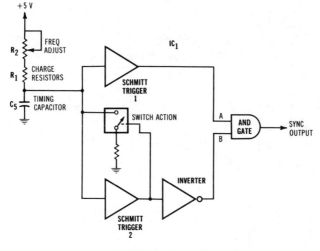

(A) Logic diagram.

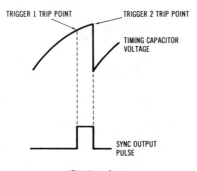

(B) Waveforms.

Fig. 14-7. Sync generator.

by side and carrying substantial currents, the technician can move away from these areas to trace current from one component to another by setting a reference current level at the node driver output pin. Then, he simply moves from pin to pin on the ICs, instead of attempting to follow along the conductors.

To estimate high and low current values with the current tracer, pulse the circuit high and low with a logic pulser. Apply the current tracer at the output of the device, and estimate the current levels from the response of the tracer. With reference to Fig. 14-10, I_{OH} represents high-level output current (logic 1); in TTL, this is approximately 40 μA. Next, I_{OL} represents low-level output current (logic 0); in TTL, this is normally about 1.6 mA

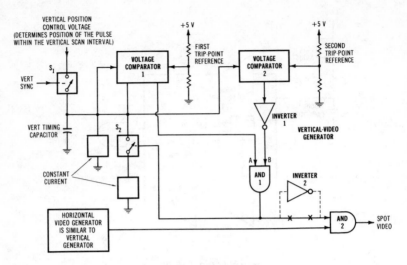

Fig. 14-8. Spot generator.

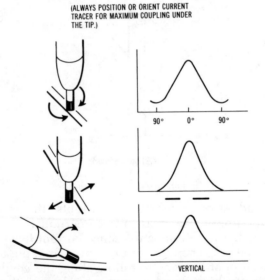

(ALWAYS POSITION OR ORIENT CURRENT TRACER FOR MAXIMUM COUPLING UNDER THE TIP.)

Fig. 14-9. Orientation of current-tracer tip. (*Courtesy Hewlett-Packard*)

for a fan-out of one (a single load). Note, however, that if the load is shorted, the TTL low-level output current may be as high as 55 mA. A partial short will result in an abnormal pulse current with a value less than 55 mA. Basic troubleshooting rules are: (1) In a known bad node, the current value usually exceeds the

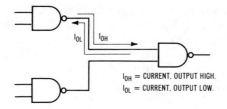

Fig. 14-10.   Output currents of logic device. (*Courtesy Hewlett-Packard*)

other currents on the pc board by a wide margin. (2) Determine the source and the sink of current; in turn, the faulty component can often be pinpointed quickly with the current tracer and logic pulser. As a practical note, the minimum feasible spacings between pc conductors for current-tracer tests at typical current levels are shown in Fig. 14-11.

MARGINAL AND INTERMITTENT CONDITIONS

Although digital ICs tend to fail catastrophically, situations are occasionally encountered in which marginal defects result in attenuated pulse output. A gate is rated by the manufacturer for a minimum voltage level which is guaranteed to be interpreted as a high at the input, and for a maximum voltage level which is guaranteed to be interpreted as a low at the input. For example, in one widely used series of gates, 2 volts is the minimum voltage level which is guaranteed to be interpreted as a high at the input; 0.8 volt is the maximum voltage which is guaranteed to be interpreted as a low at the input. This series of gates is rated for a minimum output voltage in the high state of 2.4 volts; 0.4 volt is the rated maximum output voltage in the

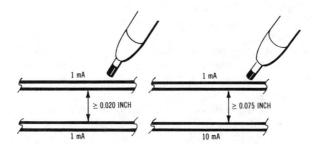

Fig. 14-11.   Minimum spacings between pc conductors for current-tracer tests. (*Courtesy Hewlett-Packard*)

311

low state. Therefore, it is good practice to observe the peak-to-peak voltage of a digital waveform when a marginal situation is suspected, and to make certain that the voltage value is within tolerance. Otherwise, a good driven gate could be confused with a marginally defective driver gate. Similarly, it is good practice to check the power-supply voltage, and to make certain that it is not subnormal. When a power-supply defect causes V_{CC} to become marginal, gates that are near their tolerance limits will appear to be defective, although they will function normally when V_{CC} is brought back to its correct value.

Digital-logic troubleshooting proceeds under three general approaches: Signal substitution, signal tracing, and aggravation. Aggravation techniques are used in troubleshooting intermittents in order to locate the area of malfunction. Thus, without the use of excessive force, the technician may proceed to twist and pull connections, cables, plugs, and plug-in units while monitoring the output of the equipment with a logic probe. It is sometimes helpful to "wipe" the handle of a plastic screwdriver across the back of a suspected row of modules to initiate an intermittent. Individual modules may be tapped to narrow down the possibilities; suspected modules can be wiggled up and down. Although most intermittents are mechanical, some are thermal. A thermal intermittent can often be initiated by blowing hot air into suspected areas, followed by a stream of cold air.

TROUBLESHOOTING MODULAR SYSTEMS

Troubleshooting of modular systems generally starts with substitution of known good boards for suspected boards. It is sometimes helpful to "wiggle" a suspected board while monitoring the output with a logic probe in order to identify possible poor contacts. An electronic thermometer is helpful in some trouble situations. When an IC is shorted internally, it dissipates more power than in normal operation, and therefore it runs hotter. If the surface temperatures of suspected ICs are checked, it may be possible to pinpoint a defective device. Note, however, that an external short, as between pc conductors, can also cause an IC and run hot because of excessive current demand.

Another troubleshooting technique is termed *card-swapping*. It is sometimes possible to get a "handle" on a malfunction by interchanging pc boards and observing the resulting outputs. The troubleshooter may evaluate card-swapping tests in either one of two ways: First, he may be guided by case histories; second, he may have an understanding in depth of system operation, whereby he can correctly interpret the significance of card-swapping. However, indiscriminate card-swapping can be a disastrous procedure; before any pc boards are interchanged, the technician should make certain that he fully understands the changes in circuit configurations and resulting signal responses that will occur. In summary, if card-swapping is done knowledgeably, useful preliminary troubleshooting data can frequently be obtained.

If a similar unit of digital equipment in normal operating condition is available, the technician has an advantageous position. The digital equipment under test and the normal reference unit are first compared with respect to their states at rest. Then, a simple routine is run on each unit, and the logic states at various test points are compared. Additional routines may be checked out on both units in a similar manner. Comparative tests are useful in "tough dog" situations because no service manual can be completely comprehensive. Comparative tests are also helpful when the technician does not have a complete understanding of system operation and must take a "shotgun" approach.

Standard documentation of digital equipment includes instruction manuals, maintenance manuals, and service indexes in addition to circuit schematics and logic documentation. Manuals and service indexes often provide facts concerning normal functions versus fault conditions. An instruction manual is primarily concerned with normal operation of the digital equipment. A maintenance manual provides timing diagrams, often with photographs of the pertinent oscilloscope patterns. A service index includes charts for troubleshooting procedures. Unless the technician is completely familiar with the equipment under test, it is essential that he consult all available documentation.

Chapter
15

Types of Memories

This chapter introduces the subject of memories, those devices and systems in which digital information is stored. Both bipolar and CMOS memory devices will be considered. Troubleshooting pointers will conclude the chapter.

SHIFT-REGISTER MEMORY

There is no sharp dividing line between shift registers and memories. When a shift register is operated as a memory, data bits are clocked in and out of the device sequentially. The various stages of the shift register are termed *memory cells*. Shift-register memories may be very elaborate and may accommodate a thousand bits of data or more. There are two basic types of shift-register memories, termed *dynamic* and *static* devices. A dynamic memory uses memory cells that store data bits in capacitive elements. Therefore, a dynamic memory must be clocked continuously at some minimum specified clock rate in order to refresh the charges in the capacitors. On the other hand, a static shift-register memory uses flip-flop cells, and continuous clocking is not required.

MEMORY MATRIX ARRANGEMENTS

An important type of memory accepts 1 and 0 bits; these bits can then be accessed and outputted as required. A *random-access memory* (RAM), also called a read-and-write memory, may have data written into it from a keyboard, for example; then, the data stored in the memory may be arbitrarily accessed and read out for performance of a chosen operation, such as addition. Another important type of memory contains 1 and 0 bits in permanent storage; it is called a *read-only memory* (ROM). Although the contents of a RAM can be erased and new data can be written into its cells, this is not possible with a ROM. A ROM is programmed, which means that data can be entered into the ROM only once; programming is commonly accomplished at the time of manufacture. As an illustration, an algorithm may be programmed into a ROM; then, whenever this algorithm is required for an arithmetic operation, it is read out of the ROM. Note also that the contents of an ordinary RAM are erased if its power supply is turned off; the RAM is called a *volatile memory*. On the other hand, a ROM is a *nonvolatile* memory.

A widely used basic type of memory consists of "rows" and "columns" of flip-flops, as shown in Fig. 15-1. The rows are usually designated as Y_1, Y_2, Y_3, Y_4, and so on, whereas the columns are designated as X_1, X_2, X_3, X_4, and so on. An address is a digital word that identifies a specific location in the memory. As an illustration, the arrangement in Fig. 15-1 has a capacity of 16 bits. In the drawing, the unshaded blocks (flip-flops) represent 0 bits, and the shaded block represents a 1 bit with the address X_3, Y_3. In case this diagram is for a RAM, it is indicated that a 1 bit has been written into the memory at X_3, Y_3, and conversely that this 1 bit can be retrieved at X_3, Y_3. Note that after this

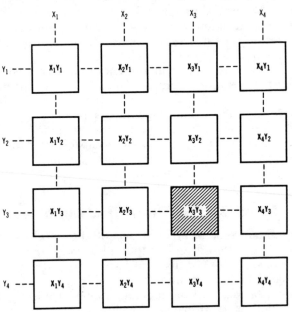

Fig. 15-1. A basic memory matrix arrangement.

information is no longer required, it can be erased, and a 0 bit may be written into the memory at X_3, Y_3. A RAM with a capacity of more than 1000 bits can be contained in a conventional IC package with 16 terminals. The position of a memory in a computer is shown in Fig. 15-2.

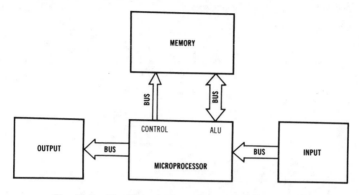

Fig. 15-2. Position of a memory in a digital computer.

MEMORY CELL

A memory cell in TTL logic for a RAM has the configuration shown in Fig. 15-3. It consists of a pair of gates connected back-to-back, thereby providing flip-flop action. To store a 1 or a 0 in a cell, the X select and Y select lines for the cell must first be addressed. The cell is then accessed because the emitters connected to the X and Y lines are driven logic-high. If a 0 is to be written into the cell, the 0 sense line is driven logic-high. If a 1 is to be written into the cell, the 1 sense line is driven logic-high. After digital data has been stored in a cell, the resulting emitter potentials indicate (sense) the logic state of the cell. In other words, if a 0 is stored, the 1 sense line will have a potential of about 1.5 V, whereas the 0 sense line will have a very low potential (the sense lines are active-low).

Because the sensing potential is in the bad region insofar as TTL action is concerned, the sensing voltage must be stepped up by buffer amplifiers in the sense lines (Fig. 15-4A). The example of a memory matrix in Fig. 15-4A is called a 16-word by 1-bit RAM; this terminology means that the matrix can store a total of 16 bits, and that each stored bit is individually accessible. Note in Fig. 15-3 that V_{cc} is applied to the collector circuit of each cell flip-flop, and that ground return is made via the X select and Y select lines.

Selection of a particular cell in a memory is called *addressing*

317

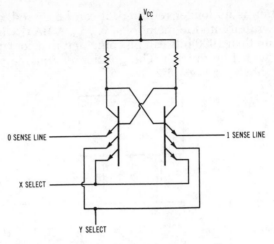

Fig. 15-3. Circuit of a memory cell.

the cell. For example, the address of the third cell from the left in the bottom row of Fig. 15-4 is X_2, Y_3. Addressing a cell does not change its state; addressing merely makes connection to the particular cell for writing in or reading out. The buffers provide circuit isolation; that is, an amplifier or an inverter is a one-way device. Thus, the inverters isolate the S_0 output from the W_0 input and the S_1 output from the W_1 input. Note that the read and write logic levels would not remain separated if buffering were not provided. A buffer has a fixed output level, regardless of the input level. Thus, a buffer may provide gain in addition to circuit isolation.

MEMORY READOUT

A memory is usually read out systematically, as illustrated in Fig. 15-5. This is an example of a 16×4 bit read-only memory (ROM). In other words, the memory contains 16 registers, and each register comprises four bits (one nibble). The registers are read out one after another. First, the address word 0000 is applied in parallel to address lines A, B, C, and D. This address becomes decoded into a high logic level on the 0 output line of the decoder. In turn, the first register is accessed, and the contents of the 0-0, 0-1, 0-2, and 0-3 memory cells appear as a four-bit binary word (nibble) on the register output lines. Next, the address word 0001 is applied to the address lines, and the contents of the 1-0, 1-1, 1-2, and 1-3 cells appear on the register

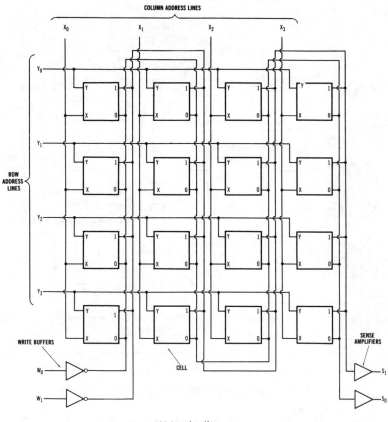

(A) Logic diagram.

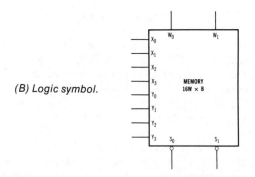

(B) Logic symbol.

Fig. 15-4.   A 16-word by 1-bit memory.

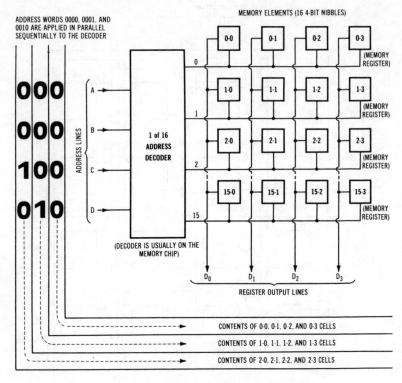

Fig. 15-5. Systematic readout of registers.

output lines. This process ordinarily continues until the entire memory has been read out.

ROM IC PACKAGE

An example of a ROM IC package is shown in Fig. 15-6. It has a capacity of 1024 bits (1 kilobit) and stores 256 words by four bits. In other words, each stored word is a nibble. The matrix is accessed by means of decoders. Each decoder translates a combination of signals into an output that represents the combination. Or, a decoder determines the meaning of an address word and initiates a memory operation whereby the information stored at the addressed location is outputted from the memory. Integrated-circuit memories usually employ on-chip decoding to minimize the number of pins required on the package. If desired, two chips can be used in tandem to double the memory capacity. Chip-select logic serves to access one chip or the other.

CMOS MEMORY

A widely used CMOS static memory-cell configuration is shown in Fig. 15-7. The eight-transistor CMOS cell in Fig. 15-8 provides for X and Y addressing, whereby the RAM can be read out bit-by-bit. Although the six-transistor CMOS cell of Fig. 15-7 is simpler, it must be read out register-by-register (word-by-word). In addition, the six-transistor CMOS cell requires a more complex decoder. On-chip decoder circuitry is usually employed, to minimize the number of package pins required. Inasmuch as this is a static arrangement, the CMOS transistors are unclocked. Data written into the cells is retained until new data may be written in. This is a volatile form of memory; if the power-supply voltage is removed, the contents of the memory will be erased.

DYNAMIC MOS MEMORY

Compared with static memories, MOS dynamic memories have the advantage of low power consumption. In turn, a dynamic memory can be fabricated in a smaller package than a comparable static memory. Because charges are stored on capacitors, a dynamic memory must be periodically refreshed to avoid loss of its contents due to charge decay; therefore, dynamic memories are clocked. A typical dynamic four-transistor MOS memory cell is shown in Fig. 15-9. To read out the cell, the address (word-select) line is driven logic-high, and the potentials of the 1 and 0 lines are sensed. To write a bit into the cell, the sense lines are driven according to the bit to be entered, and the address line is driven logic-high. To refresh the cell, its sense lines and address line are clocked. Logic states are defined in terms of negative logic.

Shown in Fig. 15-10 is a typical dynamic three-transistor MOS memory cell configuration. Note that four control lines are provided: read data, write data, read select, and write select. The read-data line is a sense line. To refresh the cell, it is clocked with alternate read and write cycles.

ACCESS TIME

Access time is also called *waiting time*. It is the time interval (read time) between the instant of calling for data from a storage

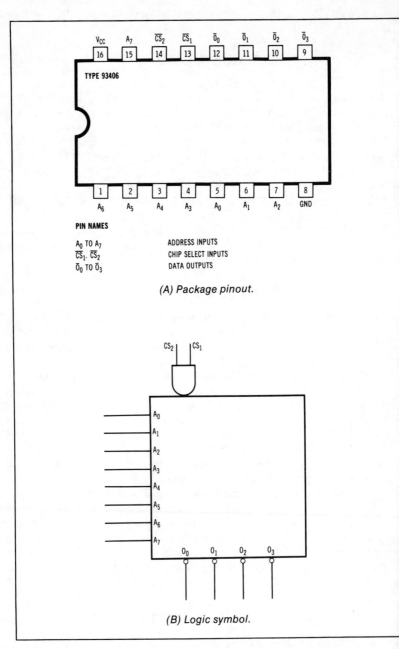

PIN NAMES

A_0 TO A_7	ADDRESS INPUTS
$\overline{CS}_1$, $\overline{CS}_2$	CHIP SELECT INPUTS
$\overline{O}_0$ TO $\overline{O}_3$	DATA OUTPUTS

(A) Package pinout.

(B) Logic symbol.

Fig. 15-6. A 1024-bit ROM IC package.

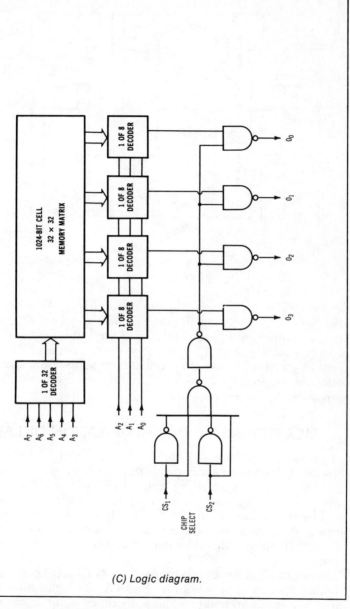

(C) Logic diagram.

(*Courtesy Fairchild Camera and Instrument Corp*.)

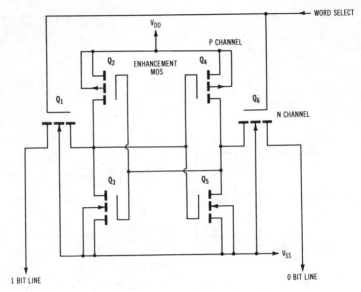

Fig. 15-7. A static CMOS memory-cell configuration.

device and the instant of completion of delivery of the data. It is also defined as the time interval (write time) between the instant of requesting storage of data and the instant of completion of storage of the data. In a memory system, access time is defined as the time delay, at specified thresholds, from the application of an enable or address pulse until the arrival of the memory data output. (See Fig. 15-11.)

MEASUREMENT OF PROPAGATION DELAY

Propagation delay denotes the elapsed time between a change of input state and a resulting change in output state. If a gate has a propagation delay time of 15 nanoseconds (ns), four gates operated in series will have a propagation delay time of 60 ns. The read access time of a RAM is typically 60 ns. Marginal devices in a digital network can cause abnormal propagation delay, with resulting development of glitches or data-processing errors.

Propagation delay time is measured with a triggered-sweep oscilloscope, preferably with A+B and A−B display modes. The input of a configuration (such as a divide-by-eight arrangement)

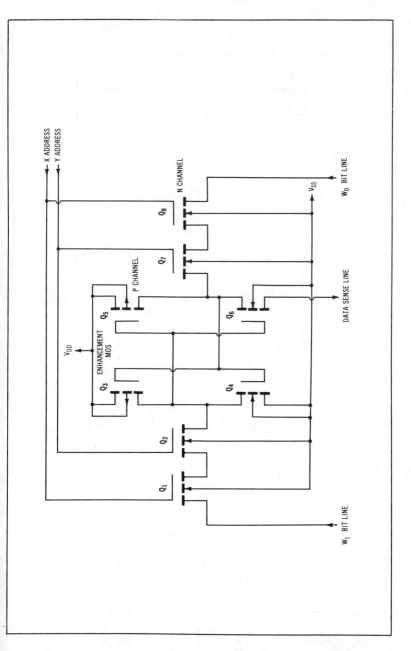

Fig. 15-8. An eight-transistor CMOS memory-cell configuration.

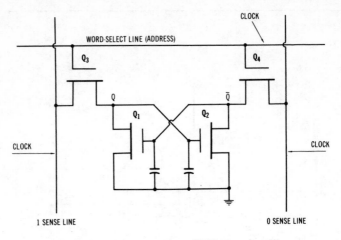

Fig. 15-9. A four-transistor MOS memory cell.

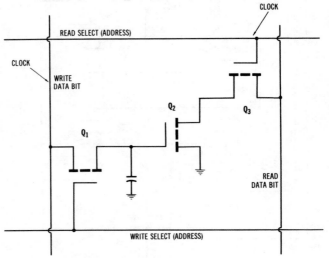

Fig. 15-10. A three-transistor MOS memory cell.

is connected to channel A of the scope, and the output is connected to channel B. The A and B signals are displayed conventionally, and the scope is then switched to the A+B or A−B operating mode. With reference to Fig. 15-12, the propagation delay, T_p, is displayed as a pulse in the A−B display. When this pulse is expanded, the elapsed time (propagation delay) indicated by the pulse can be precisely measured. Unless the documentation of the digital system specifies the normal propaga-

326

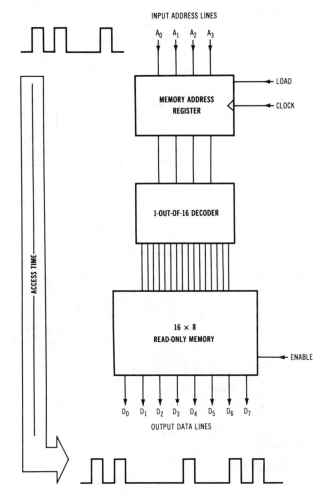

Fig. 15-11. Example of access time.

tion delay, a comparison test should be made with respect to a similar digital system that is in normal operating condition.

TROUBLESHOOTING OF MEMORY SYSTEMS

A logic comparator is the best digital instrument for memory testing. A known-good memory IC package of the same type is inserted into the comparator, and the termination of the test cable is pushed down over the IC package under test. The ter-

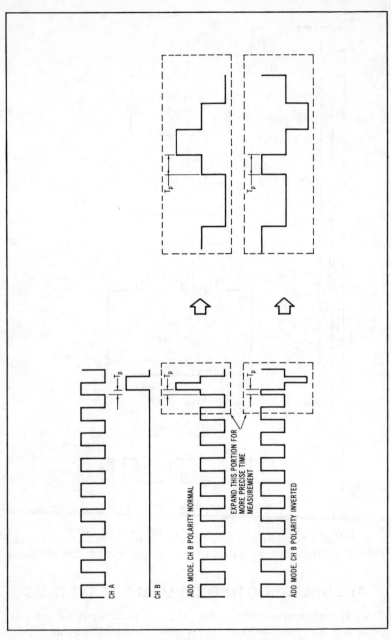

Fig. 15-12.   Measurement of propagation delay.
(*Courtesy B&K Precision*)

Chart 15-1. Comparator Capabilities Chart
(16 or Less Pin Dual-in-Line DTL, TTL)

Combinatorial Logic (AND, NAND, XOR, etc.)	Excellent. This also includes expandable ICs. This category includes the vast majority of ICs in use.
Sequential Logic (Flip-Flops)	Excellent. Reference and test IC should be synchronized by a pulse on the "Reset" input.
Memories, Shift Registers	Excellent. Clip must be attached long enough for reference and test IC to contain the same information.
Low-Power TTL	Good. The comparator is an input load of 5 for the driving device.
One-Shots	Usually good. Since reference and test IC share the RC timing components, circuit timing can be affected.
Open-Collector and Three-State Logic	Usually poor. When outputs are bused together, a good gate is constrained to operate improperly, and this will be indicated by the comparator.
Expanders, Analog and Linear ICs	No. Their outputs are analog and cannot be tested by the comparator.
MOS Devices	No. They require different power supplies exceeding the 7-V input limit and will damage the comparator.

Courtesy Hewlett-Packard

mination automatically makes contact with the pins of the IC under test. The equipment is then turned on as for normal operation. As the memory is progressively loaded and then unloaded, no discrepancy between the reference IC and the IC under test will normally be indicated by the comparator. In other words, no LED will normally glow. However, if a LED does glow, its location in the top of the comparator identifies the pin on the IC package which has produced an erroneous output. A complete checkout requires sufficient time for the memory to become fully loaded and to become fully unloaded. Note that if an error is indicated, the fault might be inside the IC under test, but it might also be situated along an input or output conductor on the pc board. A summary of logic-comparator capabilities is given in Chart 15-1.

TROUBLESHOOTING NOTES

It is sometimes feasible to identify a defective IC by means of voltage measurements. First, measure the V_{CC} supply voltage. Then measure the voltages in the circuit(s) suspected of malfunctioning. High-level voltage values in digital networks are never quite as high as V_{CC} (in normal operation). For example, in a TTL circuit with $V_{CC}=5$ V, high levels through the circuit normally range from 3.5 to 4.5 V. If a heavy load is being driven, the high level might be down to 2.4 V. Thus, if a high level equal to V_{CC} is measured, look for a short-circuit between the output pin and V_{CC}. Low-level voltages in digital networks are never zero (in normal operation). Unless a few hundred millivolts is measured at a low output, look for a short between the output pin and ground.

Sometimes pin-lifting tests are needed. Pin-lifting consists of disconnecting an IC from its socket during test procedures. In other words, an IC is unplugged from its socket, and a particular pin is bent aside; the IC is then replaced in its socket. Long-nose pliers are generally used to bend IC pins; a 45° bend is adequate. Then, the circuit action is checked with a logic pulser and logic probe, taking into account the lifted pin(s). In various situations, test results that formerly did not "make sense" may become clarified when IC pins are lifted. For example, glitches are sometimes fed back to a preset pin or a clear pin by a defective IC in another part of the circuit; the fed-back glitches can greatly confuse a trouble analysis. Remember that in some cases it may be permissible to "float" a lifted IC pin, but in other cases a lifted IC pin may have to be tied to a fixed high-level or low-level source in order to obtain a valid test of circuit action.

LOCALIZING OPEN CIRCUITS

An open signal path, such as a break in a pc conductor, will not be indicated as a fault by a logic comparator used to test a memory (or other IC). The IC will pass a comparator test, but the circuit action is incorrect. When the IC is established as normal and an open signal path is suspected, a follow-up test should be made with a logic pulser and probe to determine the presence or absence of continuity. If a point of discontinuity is present, the logic probe will pinpoint the fault.

TROUBLESHOOTING LEVELS

As shown in Fig. 15-13, there are three basic digital troubleshooting levels. Isolation of a malfunction to a functional system block is accomplished to best advantage with a logic

Table 15-1. Troubleshooting Instruments Used for Various Faults

TYPICAL DIGITAL IC TROUBLESHOOTING PROBLEM	STIMULUS/RESPONSE INSTRUMENTS	
	STIMULUS	RESPONSE
Shorted IC Input	Pulser	Current Tracer and Logic Probe
Stuck Data Bus	Pulser	Logic Probe and Current Tracer
Internal Open in IC	Pulser	Logic Probe
Solder Bridge	Pulser	Current Tracer
Sequential Logic Fault	Pulser	Logic Clip
Shorts to V_{cc} or ground	Pulser	Logic Probe

Courtesy Hewlett-Packard

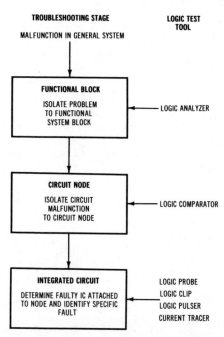

Fig. 15-13.   Three levels of digital troubleshooting.
(*Courtesy Hewlett-Packard*)

331

state analyzer. Isolation of a circuit malfunction to a circuit node is usually facilitated by the use of a logic comparator. To determine the faulty IC connected to a node, and to identify the specific fault, a logic probe, logic clip, logic pulser, current tracer, or combinations of these instruments are recommended (Table 15-1). Digital IC failures usually involve a shorted IC input, a stuck data bus, an internal open in an IC, a solder bridge, a sequential logic fault, or a short to V_{CC} or ground. A high-performance triggered-sweep oscilloscope is useful for analysis of glitches and propagation delays.

Chapter
16

Additional Memories

This chapter deals at length with the subject of programmable read-only memories. There are also discussions of character generators, memory banks, and bubble memory. For completeness, the now-obsolete core memory is also discussed briefly.

PROGRAMMABLE READ-ONLY MEMORY

A *programmable read-only memory* (PROM) is not programmed at the time of manufacture; instead, it is programmed by the user. The conventional type of PROM includes fusible emitter resistors associated with transistors in the cells of the memory, as depicted in Fig. 16-1. An intact emitter resistor in a cell corresponds to a 0 programmed in that cell (Fig. 16-2B). Conversely, a fused (blown) emitter resistor in a cell corresponds to a 1 programmed in that cell (Fig. 16-2C).

An abnormal current is applied to a particular cell in order to program a 1 into the cell. After a conventional PROM has been programmed, it is equivalent to a ROM, and its stored data cannot be changed. However, an *erasable programmable read-only memory* (EPROM) can be reprogrammed when desired. In one widely used design of EPROM, the stored data is erased by exposing the chip to ultraviolet light. Then, new data may be programmed into the EPROM. Note that an EPROM does not contain bipolar transistors; instead, field-effect transistors are utilized.

Programming Procedure

A bipolar PROM with a capacity of 512 words by 8 bits is depicted in Fig. 16-3. Four memory-enable inputs are provided to control the output states. When E_1 and E_2 are driven logic-

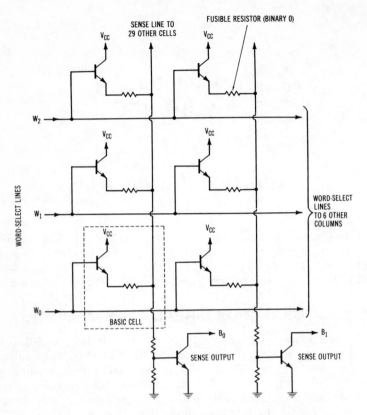

Fig. 16-1. Part of an unprogrammed 32-word by 8-bit PROM.

low and E_3 and E_4 are driven logic-high, the output presents the contents of the selected word. If E_1 and E_2 are driven logic-high or E_3 or E_4 is driven logic-low, all eight outputs go to the off, or high-impedance, state. The device is fabricated with all locations low; a high may be programmed into any selected location by means of the procedure explained below. This type of PROM, once programmed, cannot be reprogrammed.

Programming is accomplished at an ambient temperature between 15°C and 30°C. Address and chip-enable pins must be driven with normal TTL logic levels during both programming and verification. Programming occurs at a selected address when V_{CC} is held at 10.5 V and the chip is subsequently enabled. First, the desired word is selected by applying high and low levels to the appropriate address inputs. The chip is disabled by applying appropriate levels to the enable inputs.

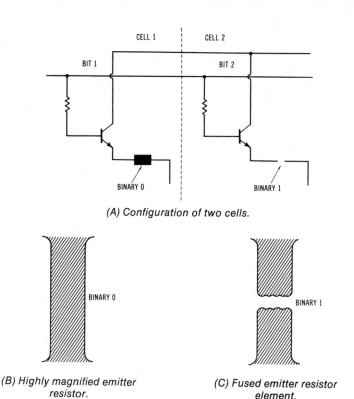

(A) Configuration of two cells.

(B) Highly magnified emitter resistor.

(C) Fused emitter resistor element.

Fig. 16-2. Details of PROM cell.

Then, V_{CC} is increased to 10.5 V ±0.5 V, with the rate of increase being between 1.0 and 10.0 V/μs. Since V_{CC} supplies the current to program the fuse as well as the I_{CC} of the device at programming voltage, it must be capable of supplying 400 mA at 11.0 V.

Next, the output at which a high level is desired is selected by raising that output voltage to 10.5 V ±0.5 V. The rate of voltage increase is limited to a value between 1.0 and 10.0 V/μs. This voltage change may occur simultaneously with the increase in V_{CC}, but it must not precede it. It is critical that only one output at a time be programmed, since the internal circuits can only supply programming current to one bit at a time. Outputs not being programmed must be left open or tied to a high-impedance source of at least 20 kilohms. The outputs of the PROM are still disabled at this time.

At this point, the device is enabled by applying appropriate levels to the chip-enable inputs. This is done with a 10-μs

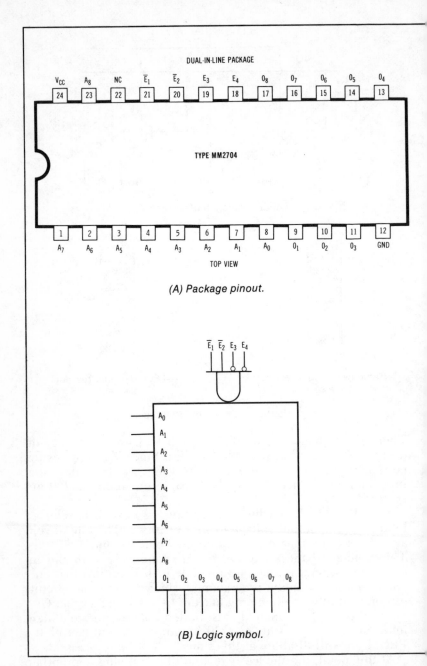

(A) Package pinout.

(B) Logic symbol.

Fig. 16-3.   A 512-word by 8-bit bipolar PROM.

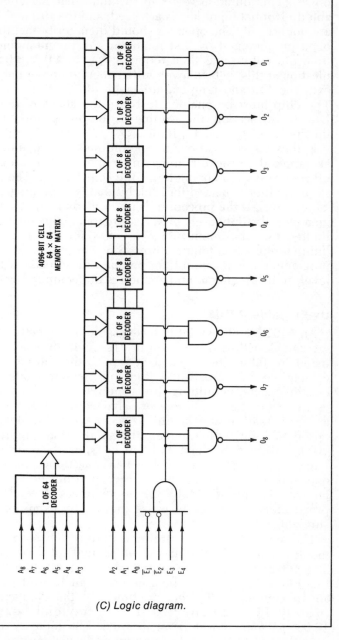

(C) Logic diagram.

(*Courtesy National Semiconductor Corp.*)

pulse. (This duration refers to the time that the circuit is enabled.) Normal input levels are used, and the rise and fall times are not critical. The operator should then verify that the bit has been programmed by first removing the programming voltage from the output and then reducing V_{CC} to 4.0 V ± 0.2 V. Verification at this voltage level will guarantee proper output states over the V_{CC} and temperature range of the programmed part. The chip must be enabled to sense the states of the outputs. During verification, the loading of the output must be within the specified I_{OL} and I_{OH} limits.

Following verification, five additional programming pulses are applied to the bit being programmed. The programming procedure is then complete for the selected bit. The foregoing steps are then repeated for each bit that is to be programmed to a high level. If the procedure is performed on an automatic programmer, the duty cycle of V_{CC} at programming voltage must be limited to a maximum of 25%. This limitation is required to limit chip junction temperatures. Since only an enabled chip is programmed, it is possible to program these parts at the board level, if all programming parameters are complied with.

UV-Erasable PROM

An 8192-bit ultraviolet-light–erasable and electrically reprogrammable EPROM is shown in Fig. 16-4. Pin connections are listed in Table 16-1. The data inputs and outputs are TTL-compatible during both the read and program modes. Outputs are three-state, permitting direct interfacing with common system bus structures. The erasure characteristics of this memory are such that erasure begins to occur when the device is exposed to light with wavelengths shorter than approximately 4000 angstroms. Sunlight and certain types of fluorescent lamps have wavelengths in the 3000–4000 angstrom range. Constant exposure to room-level fluorescent lighting could erase the memory in approximately 3 years, whereas it would take approximately one week to cause erasure by exposure to direct sunlight.

The recommended erasure procedure is to expose the device to ultraviolet light that has a wavelength of 2537 angstroms. The integrated dose (i.e., ultraviolet intensity times exposure time) for erasure should be a minimum of 15 watt-seconds per square centimeter. The erasure time with this dosage is approximately 15 to 20 minutes if an ultraviolet lamp with a 12,000-μW/cm^2 power rating is used. The memory should be placed within 1 inch of the lamp tubes during erasure. Note that

338

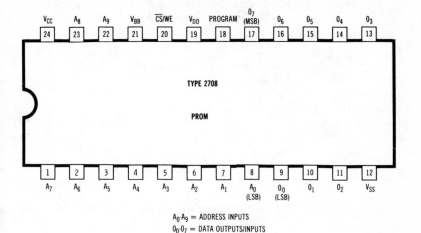

A_0-A_9 = ADDRESS INPUTS
0_0-0_7 = DATA OUTPUTS/INPUTS
$\overline{CS}$/WE = CHIP-SELECT/WRITE-ENABLE INPUT

(A) Package pinout.

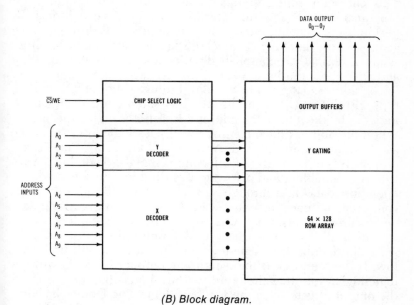

(B) Block diagram.

Fig. 16-4.   An ultraviolet-erasable electrically reprogrammable EPROM.
(Courtesy Intel Corp.)

Table 16-1. Pin Connections During Read or Program for Type 2708 EPROM

Mode	Data I/O (Pins 9-11, 13-17)	Address Inputs (Pins 1-8, 22, 23)	V_{SS} (Pin 12)	Program (Pin 18)	V_{DD} (Pin 19)	$\overline{CS}/WE$ (Pin 20)	V_{BB} (Pin 21)	V_{CC} (Pin 24)
Read	D_{OUT}	A_{IN}	Gnd	Gnd	+12 V	V_{IL}	−5 V	+5 V
Deselect	High Impedance	"Don't Care"	Gnd	Gnd	+12 V	V_{IH}	−5 V	+5 V
Program	D_{IN}	A_{IN}	Gnd	Pulsed 26 V	+12 V	V_{IHW}	−5 V	+5 V

Courtesy Intel Corp.

some lamps have a filter on their tubes; the filter should be removed before the erasure process is carried out.

CHARACTER GENERATOR

The device in Fig. 16-5 is a widely used bipolar character generator with serial output, designed primarily for crt display. This character generator incorporates several crt system-level functions, as well as a character font in a 7×9 matrix format. It performs the system functions of parallel to serial shifting, character address latching, character spacing, and character line spacing (normally accomplished with extra packages). Figs. 16-6 and 16-7 show the basic plan of this character generator.

The address latches are the "fall-through" or "feed-through" type (Fig. 16-8). When the address latch-control signal is logic-high, the character addresses "fall through" the latch. When the address latch-control signal goes logic-low, the character addresses are latched. A 40-ns address setup time is required. When the address latch-control signal is logic-high, AND gate A_1 is enabled, and AND gate 2 is disabled. In this mode of operation, data "falls through" the latch. When the address latch control signal goes logic-low, gate A_1 is disabled, blocking any new address inputs. Gate A_2 is enabled by a logic-high on input A, which allows the feedback to determine the output of gate A_2. If the feedback is logic-low, the output of A_2 will be logic-low. If the feedback is logic-high, the output of gate A_2 will be logic-high. Note that there are two inversions (O_1 and I_1) from the output of gate A_2 to the feedback loop. The feedback maintains the level that was present on inverter I_1 when the address latch control became logic-low.

The ROM capacity is $64\times7\times9 = 4032$ bits. The National

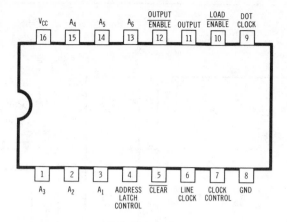

(A) Package pinout.

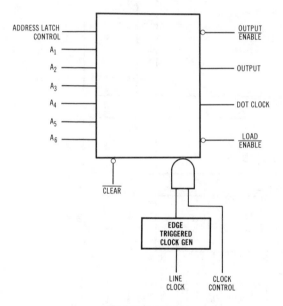

(B) Logic symbol.

Fig. 16-5.   Type DM8678 character generator. (*Courtesy National Semiconductor Corp.*)

Semiconductor ROM comes with a standard upper-case character set. However, other fonts may be utilized, if desired.

The input clock is shaped by an edge-triggered clock generator (Fig. 16-6). The output clock pulse from the generator

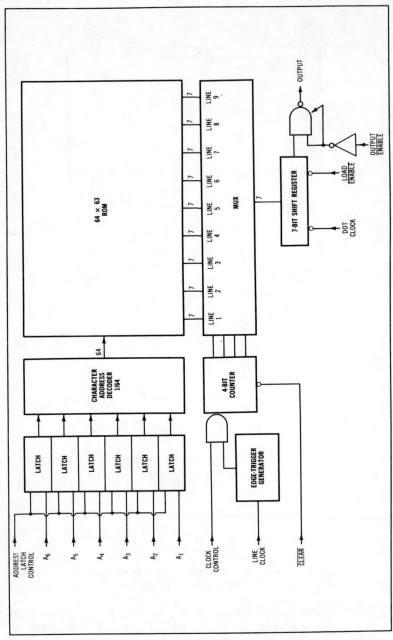

Fig. 16-6.   Block diagram of Type DM8678 character generator.
(*Courtesy National Semiconductor Corp.*)

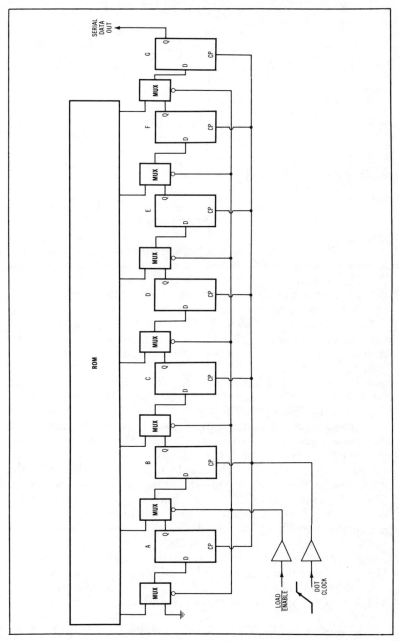

Fig. 16-7.   Logic diagram of Type DM8678 character generator.
(*Courtesy National Semiconductor Corp*.)

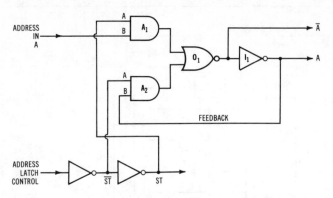

Fig. 16-8. Address latch for character generator. (*Courtesy National Semiconductor Corp.*)

is enabled by the clock control signal; the output pulse from the clock generator goes to one input of a two-input AND gate, and the clock control signal goes to the other input of the AND gate. When the clock-control signal is logic-low, the clock signal is blocked by the AND gate.

The line counter is shown in Fig. 16-9. It is a mod-16 counter, and its count can be shortened by clear, which resets the counter to its first state when it goes to the logic-low level.

A seven-bit parallel-in/serial-out shift register is used to serialize the output state. Seven D-type flip-flops and seven 2-line–to–1-line multiplexers are used to perform the parallel-to-serial conversion, as shown in Fig. 16-7. Operation of the parallel-to-serial converter starts with a cycle which begins with the load-enable going logic-low. This condition routes data from the ROM via the multiplexers (MUX) to the D inputs of the seven flip-flops. The data at the D inputs is clocked into the flip-flops on the next low-to-high transition of the dot clock. Next, the load-enable goes logic-high, thereby switching the multiplexers. Now, data at each D input comes from the Q output of the preceding flip-flop stage.

The first stage in the shift register is an exception to the foregoing action; the multiplexer routes a logic-low to the first D input. After seven clock cycles, all stages are logic-low, and any additional clock cycles will produce a logic-low output. This characteristic is used for horizontal spacing between characters.

The output buffer (Fig. 16-6) is a three-state output circuit. When the output enable is logic-high, the output is in the high-impedance state. The output can sink 16 mA at 0.45 V for a

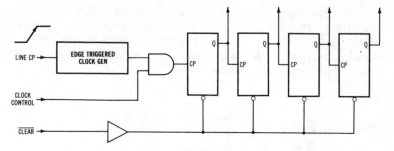

Fig. 16-9. Line counter for character generator. (*Courtesy National Semiconductor Corp.*)

logic-low output signal, and it can source 2 mA at 2.4 V for a logic-high output signal.

To illustrate the operation of the character generator, the following example gives the sequence of events involved in generating the character "N." Generation begins with the appropriate six-bit character address becoming valid on address inputs A_1 to A_6 (Fig. 16-6). This address can be latched by bringing the address latch-control signal logic-low. There are address and hold times of 50 ns and 40 ns, respectively. The output from the address latch is decoded in the character address decoder. This is a 1/64 active-high decoder. The word line which contains the code goes logic-high.

The ROM contains the codes required for generating the 64 7×9 characters. The ROM is organized with 64 words, each 63 bits (7×9=63) in length. In the ROM, the first seven bits of 63 are line 1 of the character. The next seven bits store line 2 of the character, and so on. The lines and dots in this example are illustrated in Fig. 16-10. The code for N would be:

1000001	1000001	1100001	. . . . . .
Line 1	Line 2	Line 3	
1000011	1000001	1000001	
Line 7	Line 8	Line 9	

After the access time has elapsed (tas = 350 ns), the output from the ROM for line 1 of the character N can be loaded into the shift register. When load-enable is brought logic-low, the shift register is loaded with line 1 of the character N on the next low-to-high transition of the dot clock. The next six dot-clock cycles shift out the rest of the first line of character N. If only a single character in a row was generated, the line clock would go from low to high, advancing the line counter, which in turn

345

switches the multiplexer to line 2 of the character. Line 2 contains the next seven bits required for generating N. This would continue until the ninth line has been clocked out. Any additional line-clock cycles will put a vertical space between characters. This action is illustrated in Fig. 16-10.

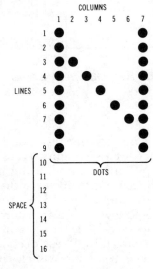

Fig. 16-10. Example of dot pattern for letter "N." (*Courtesy National Semiconductor Corp.*)

In a typical application, more than one character is displayed in a row, as shown in Fig. 16-11. The sequence is as follows: Line 1 of the first character is clocked out. Note that seven dot-clock cycles are required to shift out one line in a character. Additional dot-clock cycles add logic-lows to the end of the line. This action provides a horizontal space between characters. There is no limit to the number of clock cycles that can be used to generate horizontal spacing. The address is changed to select the second character. Then the first line of the second character is clocked out. Next, the first line of the third character is clocked out, and so on until the first line of the last character in the row has been clocked out. At this time, the line counter is clocked, thereby advancing the line counter to line 2. The first character is again addressed, and the process of scanning continues until the ninth line of the last character in the first row has been shifted out. Thereupon, the line clock is again activated, thereby incrementing the line counter to line 10. All characters in the row have been scanned, so the output from the character generator for lines 10 to 16 is all logic lows. This provides a vertical space between rows. The number of lines that are used to space can be controlled by making clear logic low

346

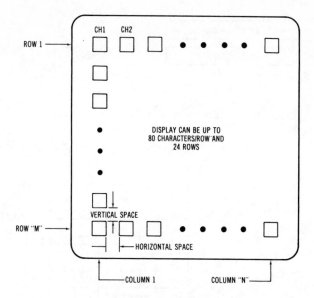

ROW 1

CH1 CH2

DISPLAY CAN BE UP TO
80 CHARACTERS/ROW AND
24 ROWS

VERTICAL SPACE

ROW "M"

HORIZONTAL SPACE

COLUMN 1

COLUMN "N"

Fig. 16-11. Example of crt display. (*Courtesy National Semiconductor Corp.*)

after the desired number of lines of vertical space have been generated. Then, the first line of the first character of the second row is addressed and scanned. The process continues until the ninth line plus lines for vertical spacing of the last character in the last row have been scanned. At this point, the display field has been written. For a crt display, it is necessary to refresh the display. Displays are typically refreshed 30 to 60 times each second. The memory is required to store the character addresses so that they may be called up when required for refresh.

MEMORY BANK

A *memory bank* or *array* is an arrangement of cells in three dimensions, as illustrated in Fig. 16-12. Cells are grouped in *planes*, also called *cards* or *chips*. Each plane contains corresponding bits of each stored word. Thus, the bank in Fig. 16-12 accommodates eight-bit words. Each plane has a capacity of eight eight-bit words, and the eight planes (bank) have a capacity of 512 bits (64 eight-bit words). This is an example of a read/write memory (RAM). The matrix size denotes the number of words that can be stored. The bank in Fig. 16-12 has planes that consist of memory chips; each plane comprises four 16-bit

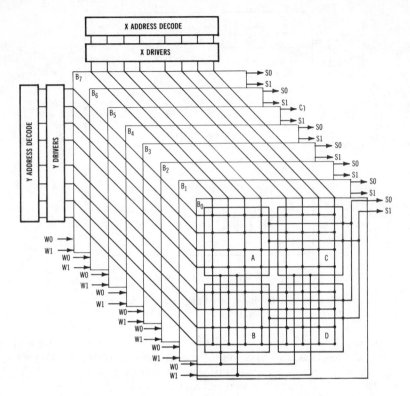

Fig. 16-12. Arrangement of an eight-plane memory. (*Courtesy Texas Instruments Inc.*)

memory chips, connected into an 8×8 matrix. The matrix is completed by paralleling the X lines of circuits AB and CD, and by paralleling the Y lines of circuits AC and BD to form the 64 addresses of the B_0 plane. Similar configurations are provided for the seven other planes.

CORE MEMORY

Core memory is an obsolescent type of RAM wherein the basic storage device is a small toroidal form of magnetic material (Fig. 16-13). Each toroid can be magnetized in one direction to represent a 0, or it can be magnetized in the opposite direction to represent a 1. This is a nonvolatile memory, since if the power is removed, the stored information remains. On the other hand, a core memory is subject to destructive readout, so that circuitry must be provided for rewriting the stored data auto-

348

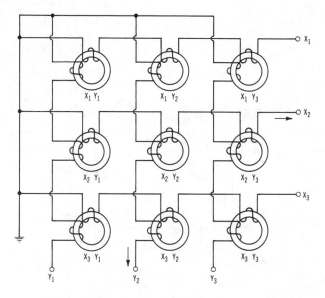

Fig. 16-13. Arrangement of a core memory.

matically. Access time is comparatively slow, compared with semiconductor memories. A *core plane* is a horizontal network of magnetic cores that contains a core that is common to each storage position.

BUBBLE MEMORY

A *bubble memory* utilizes a magnetic substance in which a very small circular domain of magnetization, called a *bubble,* is stored to represent a logic 1 bit (Fig. 16-14). Magnetic garnet with single-crystal thickness is used; amorphous films are also employed. Application of an electrical pulse serves to store a bubble. The bubble is constrained to a circular shape by means of an external magnetic bias field. A bias field is also used to move a bubble from one location to another in the magnetic substance. Nondestructive readout of stored data is obtained by converting the magnetic field of a bubble back into an electrical pulse. Various combinatorial logic operations can also be performed with the stored bubbles.

Bubbles are moved along paths that are determined by Permalloy patterns deposited on an oxide layer that covers the garnet. Motion is caused by a rotating magnetic field. One-megabit bubble-memory chips are commercially available. The average

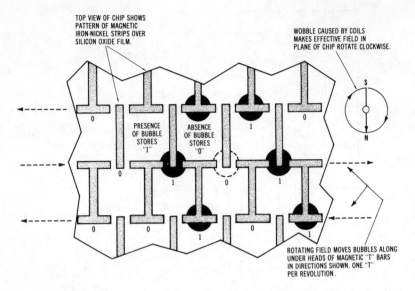

Fig. 16-14. Plan of a bubble memory. (*Courtesy Texas Instruments Inc.*)

access time is 6 ms. Bubbles are approximately 2 μm in diameter. A typical data rate is 500 kb/s. Bubble memories are nonvolatile. Architecture is basically shift-register oriented in the present state of the art; random access has not been achieved. The chief advantage of a bubble memory is its large storage capacity.

Chapter
17

Miscellaneous Bipolar
and MOS Memories

Semiconductor memories are categorized into three broad groups (Fig. 17-1). These are the *random-access, read-only*, and *serial* types. With two exceptions, each of these three categories can be implemented either with MOS or bipolar semiconductor families. Note, however, that charge-coupled devices (CCDs) and erasable programmable ROMs (EPROMs) can be implemented only in MOS technology.

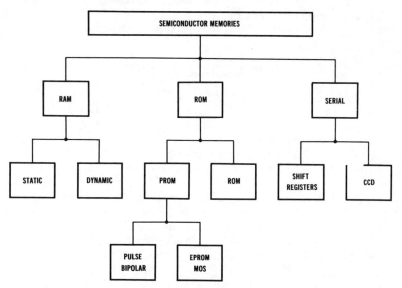

Fig. 17-1.   Classes and subclasses of semiconductor memories.

BUCKET BRIGADE

The bucket brigade in Fig. 17-2 is a shift register that transfers charges from stage to stage in response to a clock signal; it is a charge-coupled device. In a widely used application, a bucket brigade is used to delay an audio signal. The input stage operates as an analog-to-digital converter; instantaneous amplitudes of the audio signal are sampled at rapidly successive intervals and converted into appropriately coded digital pulses. The chain of digital pulses is then passed along a bucket brigade, such as the one in Fig. 17-2. The propagation time of the shift register provides an adjustable delay from input to output.

A typical bucket-brigade configuration uses 512 stages, with tetrode-type MOS transistors in p-channel form. Two or more integrated circuits may be cascaded for time delays greater than 25 ms. The signal-delay time is related to the clock frequency as shown in Fig. 17-3. Frequency response is also a function of clock frequency. For example, a single bucket brigade operating at a clock frequency of 10 kHz has a delay time of approximately 25 ms. If the clock frequency is 500 kHz, the delay time is 0.5 ms. A clock frequency of 100 kHz provides an audio bandwidth of 10 kHz; however, a clock frequency of 40 kHz provides an audio bandwidth of 5 kHz; a clock frequency of 10 kHz provides an audio bandwidth of 1.5 kHz (Fig. 17-4). The bucket brigade imposes an 8.5-dB loss in the passband.

CCD SERIAL MEMORY

A *serial memory* is defined as a memory wherein data is stored in series, and reading or writing of data is accomplished in time sequence, as with a shift register. A serial memory has slow to medium access time. A 16,384-bit charge-coupled device (CCD) serial memory is shown in Fig. 17-5. It comprises 64 recirculating shift registers of 256 bits each. Note that *latency time* in serial storage operation is the time that is necessary for the desired storage location to appear; it is analogous to access time. In this example, the average latency time is less than 100 μs. The maximum serial data transfer rate is 2 megabits/ second. Address registers are on-chip, as seen in Fig. 17-5B.

Charge-coupled devices in this type of memory utilize a four-phase clock that develops bucket-brigade action. The memory is a form of dynamic shift register. This memory is organized in the form of 64 independent recirculating shift regis-

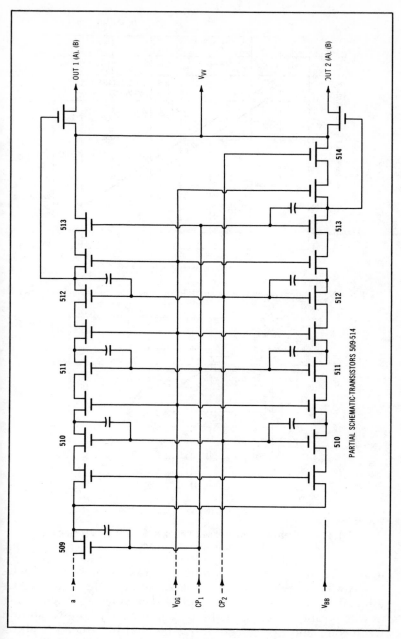

Fig. 17-2.   Arrangement of a simple bucket-brigade time-delay unit.
(*Courtesy Radio Shack*)

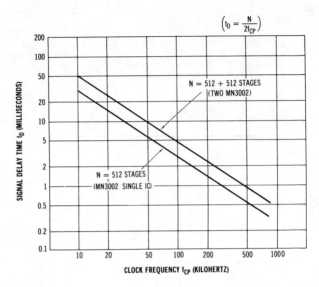

$$\left(t_D = \frac{N}{2f_{CP}}\right)$$

Fig. 17-3. Signal delay versus clock frequency. (*Courtesy Radio Shack*)

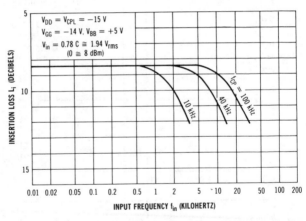

Fig. 17-4. Frequency response of bucket brigade versus clock rate.
(*Courtesy Radio Shack*)

ters to obtain a short latency time. Any one of the 64 shift registers can be accessed by applying an appropriate 6-bit address input. The shift registers recirculate data automatically as long as the four-phase CCD clocks are continuously applied and no write command is received. A one-bit shift is initiated in all 64 registers following a low-to-high transition of either ϕ_2 or ϕ_4. After the shift operation, the contents of the 64 registers at the

354

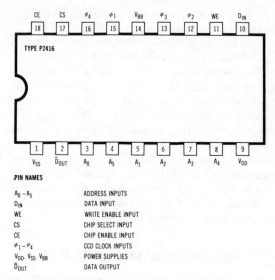

PIN NAMES

$A_0 - A_5$	ADDRESS INPUTS
D_{IN}	DATA INPUT
WE	WRITE ENABLE INPUT
CS	CHIP SELECT INPUT
CE	CHIP ENABLE INPUT
$\phi_1 - \phi_4$	CCD CLOCK INPUTS
V_{DD}, V_{SS}, V_{BB}	POWER SUPPLIES
$\overline{D}_{OUT}$	DATA OUTPUT

(A) Package pinout.

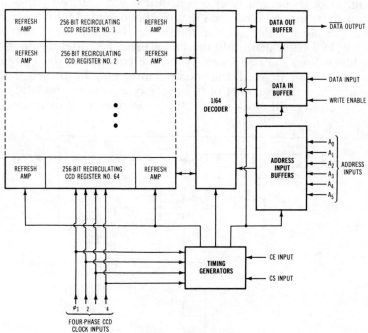

(B) Block diagram.

Fig. 17-5. A 16,384-bit CCD serial memory. (*Courtesy Intel Corp.*)

355

bit location that is involved are available for nondestructive readout and/or for modification.

Input/output functions are accomplished in a manner similar to that of a 64-bit dynamic RAM. At the next shift cycle, the contents of the 64 accessible bits (whether modified or not) are transferred forward into the respective registers, and the contents of the next bit of each register become accessible. No i/o function can be performed during the shift operation. This serial memory generates and utilizes an internal reference voltage which requires some time to stabilize after the power supplies and four-phase clocks have been turned on. No i/o functions should be performed until the four-phase CCD clocks have executed at least 4000 shift cycles with the power supplies at operating voltages. After this start-up period, no special action is required to maintain the internal reference voltage stable.

RECIRCULATING MEMORY

A recirculating memory has the basic configuration shown in Fig. 17-6. Its central device is a shift register. Its output is fed back to its input so that data that has been loaded into the register will clock indefinitely through the register, out of the last stage, and back again into the first stage. The serial data output is taken from the last stage. When the contents of the memory are no longer needed, the mode control may be driven logic-high. Then the contents of the memory cannot recirculate, and the contents will clock out. At the same time, new data may be

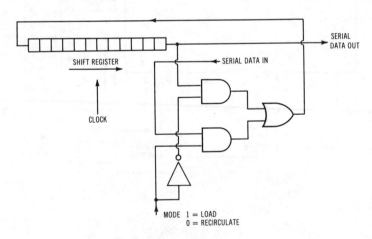

Fig. 17-6. Basic configuration of recirculating memory.

loaded via the serial data input. If the mode control is then held logic-low, the new data will recirculate indefinitely. The AND-OR gate with inverter is sometimes called a *data selector*.

A recirculating memory such as the Texas Instruments 3133 is over a thousand bits long. This implementation is termed a *mass memory*. It is a serial-access type of memory because the contents appear serially at the output. Thus, a utilization device does not have immediate access to an arbitrary stored bit; instead, the utilization device must wait for the particular bit to be clocked through to the output terminal of the memory.

MEMORY CARD

A *memory card* is a printed-circuit board on which is mounted a plurality of IC memories, usually with an edge-connector on the end of the board. Fig. 17-7 shows a diagram of a seven-unit RAM card. Each RAM has a capacity of 256 bits. Since 256 cells are to be accessed, an 8-bit address bus is utilized ($2^8=256$). All seven RAMS are addressed simultaneously. However, the particular RAM or RAMs that will be accessed are determined by the state(s) of the B input lines. For example, if the B_3 line is driven logic-high, only RAM 3 will respond to the inputted eight-bit address. Whether a read or a write operation ensues depends on the state of the read/write control line. If a read operation is commanded, data output also requires that the memory enable control line be driven logic-high. Typical card outlines are illustrated in Fig. 17-8. Cards may also be called *boards* or *modules*.

READ TIME

Read time is defined as the time that is required for a computer to locate an instruction word or data in its storage section, and to transfer it to the arithmetic unit where computations are performed. With reference to Fig. 17-9, a typical RAM section comprises line address decoders, word drivers, sense amplifiers, drivers, control logic, and the RAM itself. Read time is measured from the decoder input terminals to the data output.

WRITE TIME

Write time is defined as the time interval between the instant of requesting storage of data and the instant of completion of

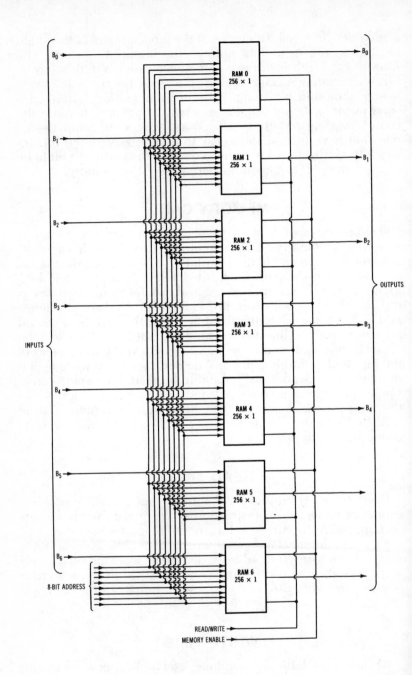

Fig. 17-7. A memory containing seven 256-bit RAMs.

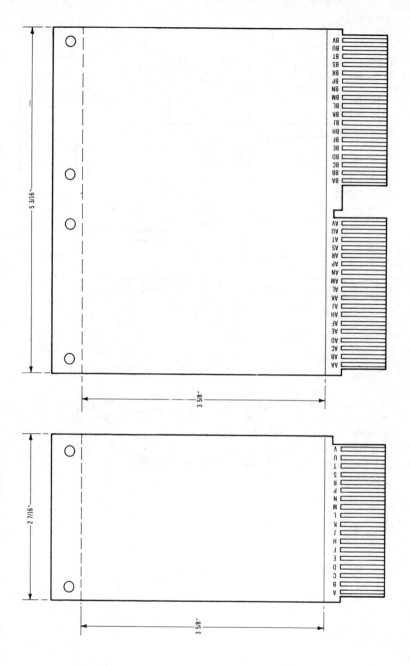

Fig. 17-8. Typical card outlines. (*Courtesy Digital Equipment Corp.*)

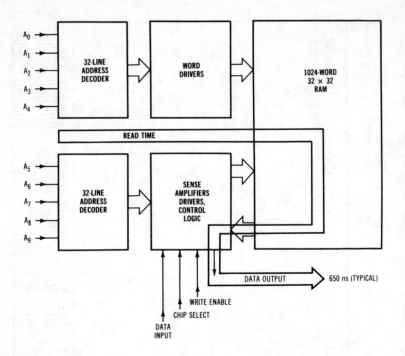

Fig. 17-9.   Example of read time for a random-access memory.

storage. With reference to Fig. 17-10, the write time is the elapsed time between application of the write-enable pulse and the termination of the write-enable pulse. Random-access memories are generally designed to have approximately equal write time and read time. A ROM has read time, but no write time.

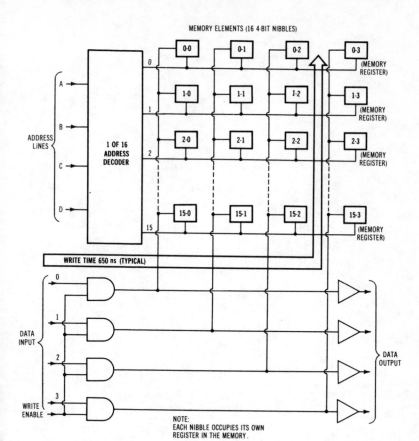

Fig. 17-10. Logic diagram showing write time for a 16-nibble RAM.

361

Chapter
18

Digital Voltmeter Circuitry

Analog-to-digital converters (a/d converters or adc's), briefly noted previously, are the "heart" of digital-voltmeter circuitry. An analog-to-digital converter is a circuit arrangement that changes a fixed or continuously varying voltage or current into a digital output. The input may be ac or dc, and the output may be serial or parallel, binary or decimal. Most digital voltmeters function to translate analog voltages into seven-segment decimal-display readout form. A typical a/d converter arrangement is shown in Fig. 18-1. It operates as follows.

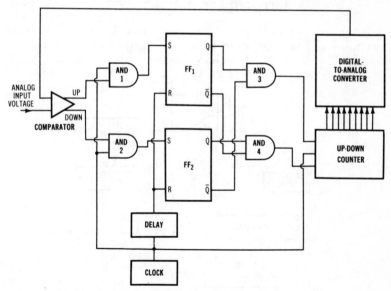

Fig. 18-1.   A typical analog-to-digital converter.

The analog input voltage is applied to a comparator that has two outputs. If the analog input voltage is greater than the digital-to-analog converter output, the *up* output from the comparator will go logic-high. On the other hand, if the analog input is less than the digital-to-analog converter (dac) output, the *down* output from the comparator will go logic-high. If the analog input voltage is greater than the dac voltage, AND gate 1 will be enabled, and the next clock pulse will set flip-flop 1; in turn, AND gate 3 will have a logic-high output, and it will enable the count-up input of the up-down counter. The counter will then count up until the dac output voltage is equal to the analog input voltage. Flip-flops 1 and 2 are reset after each clock pulse by a delayed clock pulse.

In case the analog input voltage is less than the dac output voltage, the circuit operation is essentially the same as described above, except that the count-down input of the up-down counter will be enabled. The counter will then count down until the dac output voltage is equal to the analog input voltage. Thus, the analog input voltage is changed into a binary readout provided by the up-down counter at any given clock pulse. In application, the counter output is generally applied to a seven-segment display device.

INTEGRATOR CIRCUITRY

Integrator circuitry is employed in one design of high-accuracy digital-voltmeter (dvm) circuitry. With reference to Fig. 18-2, electronic switch S_1 connects the integrator to the

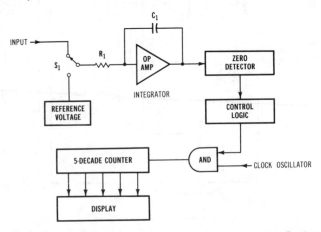

Fig. 18-2. Dual-slope integrating dvm. (*Courtesy Hewlett-Packard*)

input terminal of the dvm. This is an operational-amplifier (op-amp) integrator which is highly stable; it has a precisely linear charge-discharge characteristic. Capacitor C_1 is charged by the input voltage; then electronic switch S_1 connects the integrator to an accurate reference voltage of opposite polarity. Capacitor C_1 then discharges in a precisely linear manner through the reference-voltage source. After a certain period of time, the voltage across C_1 will equal the value of the reference voltage, and the output terminal of the integrator will fall to zero volts.

Observe that the integrator in Fig. 18-2 is followed by a zero detector with control logic that drives one input of an AND gate. The other input to the AND gate is driven by a high-precision oscillator signal. As soon as C_1 starts its discharge cycle, the control logic gates the oscillator signal through to the five-decade counter. At the instant that the integrator output reaches zero, the AND gate ceases conduction, and the counter operation stops. The counter output is decoded and displayed as a voltage value by a seven-segment display unit. This design of digital voltmeter is called a *dual-slope integrating* configuration. Nonintegrating designs are also in wide use.

LINEAR-RAMP TYPE OF DVM

A linear-ramp design of digital voltmeter is shown in Fig. 18-3. A ramp generator in the voltmeter provides a ramp voltage from a given positive to a given negative value. The input voltage is logically compared with the ramp voltage, and a count gate is opened for a period of time determined by the fall of the ramp voltage from the input-voltage value to zero. A precise oscillator voltage (clock) is passed into a counter for the foregoing period of time. As before, the counter output is decoded and displayed as a voltage value by a seven-segment display unit.

SUCCESSIVE-APPROXIMATION DVM

A successive-approximation design of digital voltmeter employs binary-coded decimal (bcd) numbers. In Fig. 18-4, the input voltage is compared to a precise reference voltage in four sequential trials (8,4,2,1). A step higher than the input voltage will be rejected by the logic, but steps lower than the input voltage are retained and are added to the voltage value from the preceding trial. The comparator has an output to its logic section, with feedback from the d-to-a converter. After the final approximation, the logic section gates the readout display. Al-

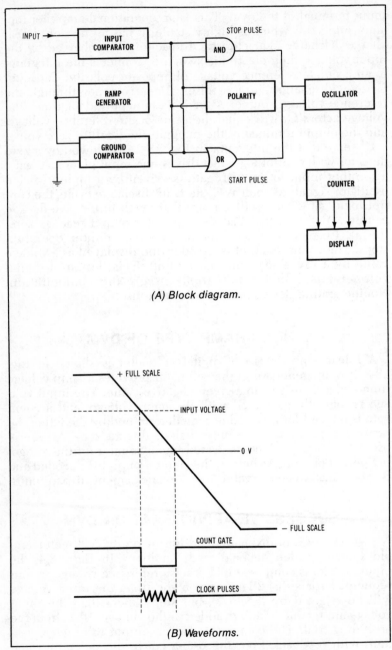

(A) Block diagram.

(B) Waveforms.

Fig. 18-3.   A linear-ramp dvm. *(Courtesy Hewlett-Packard)*

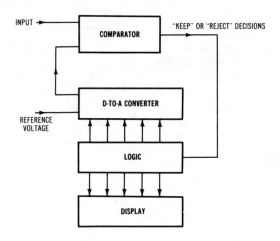

Fig. 18-4. Block diagram of a successive-approximation dvm.
(*Courtesy Hewlett-Packard*)

though the logic is clocked, the clock rate is not a critical consideration, because indication accuracy is based on the "keep" or "reject" decisions in the successive approximations.

AUTORANGING FUNCTION

Some digital voltmeters provide an autoranging function that can be used instead of the manual range-selection buttons. In a typical design, autoranging is operative on either ac- or dc-volt functions and also on the resistance function. Thus, the dvm automatically switches its instrument circuitry to the proper range of ac or dc volts or resistance. Note that a digital multimeter (dmm) provides functions such as ac voltage, resistance, dc current, and sometimes ac current indication, as well as dc voltage indication. Some of the controls on a manually operated dmm are shown in Fig. 18-5.

DIGIT DISPLAYS

Some autoranging dmm's provide a choice of 3-, 4-, or 4½-digit displays; this arrangement finds practical application in case the final digit (half-digit) "dances" in response to minor line-voltage fluctuation. A half-digit is so-called because it responds on the basis of ±50% change in the following decimal place (off-display). This type of dmm may also provide "arrow"

367

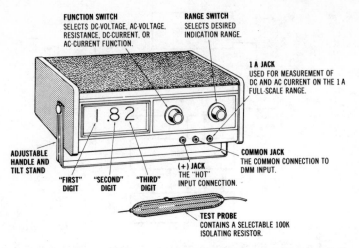

FUNCTION SWITCH
SELECTS DC-VOLTAGE, AC-VOLTAGE,
RESISTANCE, DC-CURRENT, OR
AC-CURRENT FUNCTION.

RANGE SWITCH
SELECTS DESIRED
INDICATION RANGE.

1 A JACK
USED FOR MEASUREMENT OF
DC AND AC CURRENT ON THE 1 A
FULL-SCALE RANGE.

ADJUSTABLE
HANDLE AND
TILT STAND

"FIRST" "SECOND" "THIRD"
DIGIT DIGIT DIGIT

(+) JACK
THE "HOT"
INPUT CONNECTION.

COMMON JACK
THE COMMON CONNECTION TO
DMM INPUT.

TEST PROBE
CONTAINS A SELECTABLE 100K
ISOLATING RESISTOR.

Fig. 18-5. Some typical controls on a dmm. (*Courtesy B&K Precision*)

displays for convenience in peaking or nulling tuned circuits. An "up" arrow glows if the voltage is increasing, or a "down" arrow glows if the voltage is decreasing. The more elaborate types of service dmm's also provide decibel (dBm) readout, with automatic programming for direct decibel gain comparisons.

LOW-POWER OHMMETER

Most dmm's provide a low-power ohmmeter function in addition to the conventional ohmmeter function. A low-power ohmmeter circuit operates at reduced test voltage; it applies less than 0.1 volt across the points under test. Therefore, a normal bipolar semiconductor junction cannot be "turned on," and numerous in-circuit resistance values can be measured accurately.

AC VOLTAGE FUNCTION

Various designs of ac-voltage indicating circuitry are utilized in dmm's. These range from simple half-wave ac converters (Fig. 18-6) to highly complex true-rms logic networks. The usefulness of a half-wave ac converter is limited, because correct indication is obtained only for a sine-wave input.

Some ac converters employ half-bridge or full-bridge instrument rectifiers (Fig. 18-7). However, these are also subject to

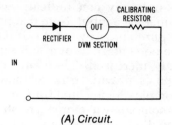

(A) Circuit.

(B) Waveforms.

Fig. 18-6. Half-wave ac converter.

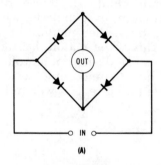

(A) Full bridge.

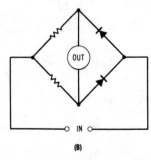

(B) Half bridge.

Fig. 18-7. Instrument rectifiers (ac converters).

369

waveform error, and they can indicate accurate ac-voltage values only for a sine-wave input. Note that precise indication is obtained only for *pure* sine-wave input; if the input sine wave has harmonics, the percentage error of ac-voltage indication will be correspondingly increased. Therefore, the more sophisticated types of dmm's provide true-rms ac-voltage indication. This requires a complex logic chip. True rms voltage values for common types of complex waveforms are shown in Fig. 18-8.

Because a silicon diode must be forward-biased to about 0.7 V and a germanium diode to about 0.3 V before effective conduction is obtained, simple ac converters are impractical for small-signal applications. However, when silicon diodes are utilized in the feedback loop of an operational amplifier, precise rectification of low-level ac signals is provided. A precision full-wave rectifier is shown in Fig. 18-9. This arrangement, or the equivalent, provides good accuracy as an ac converter.

The root-mean-square (rms) value of a waveform is based upon its power relation. Power is equal to I^2R or to E^2/R, where I or E is with reference to dc. In other words, a lamp will glow with the same brilliance, or a soldering gun will produce the same amount of heat, whether energized by 117 volts dc or by 117 rms volts ac. This is the reason that thermocouple ac converters are used in some designs of ac voltmeters. The amount of heat developed by a thermocouple depends only on the rms value of the voltage (or current) waveform.

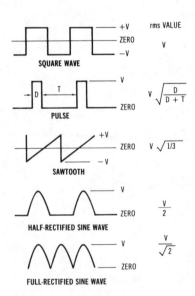

Fig. 18-8.   True rms values for common waveforms.

370

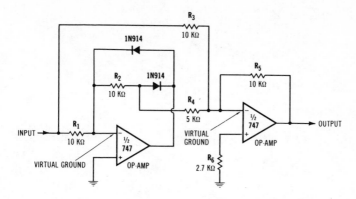

Fig. 18-9. High-accuracy ac-converter configuration.

Another ac-voltage function provided by many dvm's is peak-to-peak voltage indication. The relationship between peak-to-peak voltage and rms voltage is different for different waveforms. Since peak-to-peak voltages are universally employed in television and similar servicing procedures, it is very desirable to have a peak-to-peak function in a dvm. Peak-to-peak ac converters are very simple compared to true-rms converters. A peak-to-peak ac converter typically consists of a full-wave bridge instrument rectifier and a calibrating resistor. The output from the instrument rectifier is fed into the dc-voltage channel of the dvm.

Chapter
19

Transmission Lines in Digital Systems

A transmission line may be defined as a conductor for the transfer of electrical energy from one point to another. Thus, a cable is a transmission line. In digital systems, printed-circuit conductors are often regarded as transmission lines. That is, a conductor is analyzed as a transmission line if its length is an appreciable fraction of the operating wavelength. Fast-rise digital pulses have high-frequency harmonics; for example, a 4.5-MHz pulse has a 30th harmonic with a frequency of 135 MHz. Also, a very narrow pulse has very significant high-order harmonics. A transmission line has distributed resistance, inductance, and capacitance. It imposes more or less delay time upon a pulse from input to output. As illustrated in Fig. 19-1, a line that is open or shorted at its far end will reflect a pulse voltage (it also reflects the pulse current). The equivalent circuit for a short section of line is shown in Fig. 19-2.

Note that a uniform transmission line has a certain characteristic impedance. If a line is terminated in a resistive load that equals its characteristic impedance, a pulse will not be reflected; the pulse will be completely absorbed by the resistive load. In digital systems, however, resistive loads are the exception; most loads are more or less capacitive—a memory may present a load of 500 pF to a line. Reactive loads distort digital pulses and usually reflect a portion of the incoming pulse. Therefore, expedients are often employed in high-speed digital systems to control the amount of distortion and reflection that is developed by nonideal lines and loads.

Unless a line is terminated in a resistance equal to its characteristic impedance, the line impedance will not be constant;

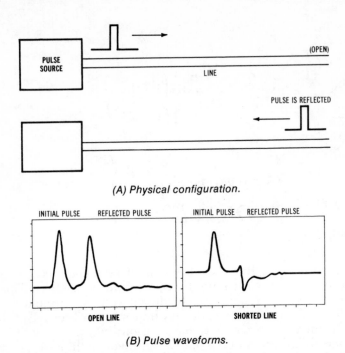

(A) Physical configuration.

(B) Pulse waveforms.

Fig. 19-1. Pulse reflection by open and shorted lines.

instead, the line impedance will vary from one point to another, as shown in Figs. 19-3 and 19-4. A line that is incorrectly terminated is called a *resonant* line, because it exhibits resonant-circuit properties. The extent of its resonant response is determined by the amount of mismatch between line and load, as well as any impedance variations (discontinuities) along the line. Observe that a 1/8-wavelength open line "looks like" a capacitor at its input terminals. A 3/8-wavelength open line "looks like" an inductor at its input terminals. A 1/8-wavelength shorted line "looks like" an inductor at its input terminals. A 3/8-wavelength shorted line "looks like" a capacitor at its input terminals. (See Figs. 19-5 and 19-6).

A quarter-wave open line "looks like" a series-resonant circuit; a half-wave open line "looks like" a parallel-resonant circuit. A quarter-wave shorted line "looks like" a parallel-resonant circuit; a half-wave shorted line "looks like" a series-resonant circuit. Any shorted line less than a quarter wavelength "looks like" more or less inductance; an open line less than a quarter wavelength "looks like" more or less capacitance. If added resistance is placed in series with a line

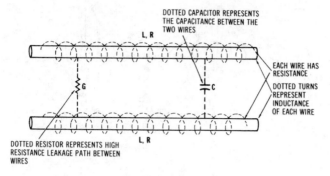

(A) Short section of two-wire line.

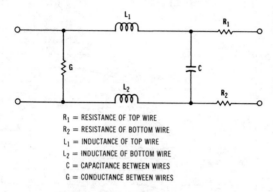

R_1 = RESISTANCE OF TOP WIRE
R_2 = RESISTANCE OF BOTTOM WIRE
L_1 = INDUCTANCE OF TOP WIRE
L_2 = INDUCTANCE OF BOTTOM WIRE
C = CAPACITANCE BETWEEN WIRES
G = CONDUCTANCE BETWEEN WIRES

(B) Equivalent circuit of line section.

Fig. 19-2.   Equivalent circuit of short section of line.

(as is commonly done in digital circuitry), the Q value of the resonant circuit is reduced. In turn, the line becomes "smoother," and reflections are reduced in amplitude. However, the added resistance wastes power, and it often operates also to slow the rise and fall of the digital pulses.

DIGITAL TRANSIENTS

Digital pulses are rectangular in theory, but they may be considerably distorted in practice (Fig. 19-7). A digital pulse may develop undershoot and ringing (Fig. 19-7B), slow rise time and slow fall time (Fig. 19-7C), or various other forms of pulse distortion. Ringing consists of a damped (decaying) sine waveform. Just prior to the ringing sequence, a downward spike occurs; this is called *undershoot*. If an upward spike occurs in a

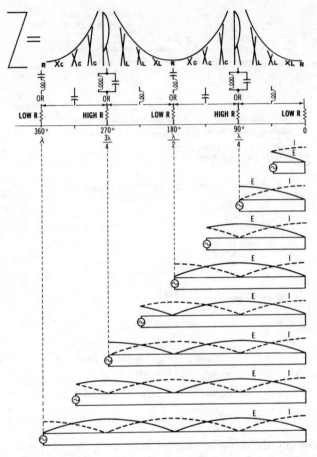

Fig. 19-3. Impedance characteristics of closed-end resonant lines.

waveform, it is called *overshoot*. Various measures are employed in digital circuitry to control the amount of waveform distortion that develops.

As indicated in Fig. 19-8, the *rise time* of a pulse is defined as the elapsed time from 10 percent to 90 percent of its final amplitude. This definition serves to separate cornering distortion from rise time. Note that *fall time* is measured in the same manner as rise time; fall time is defined as the elapsed time from 90 percent to 10 percent of the total waveform amplitude.

It is helpful to observe some of the basic causes of pulse distortion. A series RC circuit is shown in Fig. 19-9, with its resulting differentiator and integrator action on a square-wave volt-

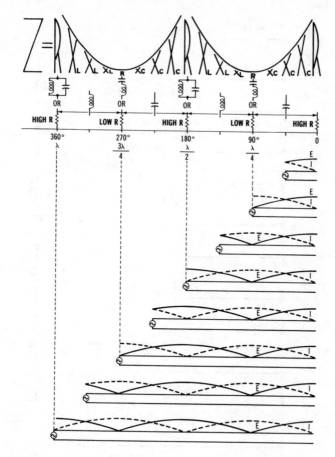

Fig. 19-4. Impedance characteristics of open-end resonant lines.

age. Note that there is a 50-V "overshoot" in the differentiated waveform, and that very slow rise time is exhibited by the integrated waveform.

The effect of the time constant on RC differentiators and integrators is shown in Fig. 19-10; the time constant is defined as the product of resistance and capacitance, or ohms times farads. The time constant (RC) is in seconds; for example, 1 megohm in series with 1 microfarad has a time constant of 1 second. If inductance and resistance are connected in series, the time constant is equal to L/R, or henries divided by ohms. Thus, 1 henry connected in series with 1 ohm has a time constant of 1 second. Note that in an LR circuit, the differentiated waveform appears

377

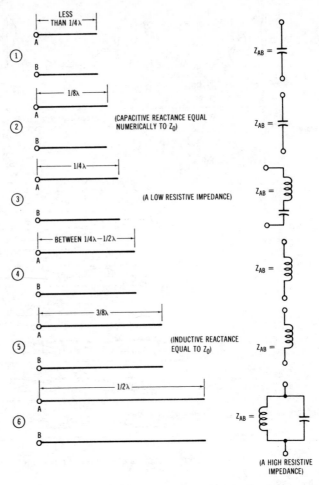

Fig. 19-5. Equivalent circuits for open-end resonant lines.

across the inductor, and the integrated waveform appears across the resistor.

The time-constant chart in Fig. 19-11 shows that the capacitor voltage in an RC circuit has the same waveform as the current in an RL circuit. Both voltage and current waveforms are of basic importance in digital circuitry.

Observe next that ringing transient response occurs when a digital pulse passes through an underdamped RLC circuit. (An underdamped circuit has comparatively low resistance.) The circuit rings at the natural resonant frequency of the L and C

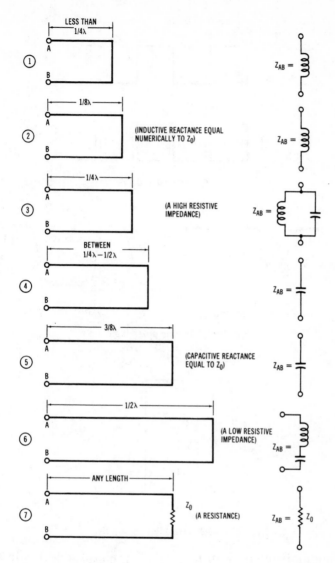

Fig. 19-6. Equivalent circuits for closed-end resonant lines.

combination; the ringing period (T) is the reciprocal of the ringing frequency (Fig. 19-12). A parallel RLC circuit will ring like a series RLC circuit; similarly, a series-parallel RLC circuit has an equivalent series circuit. Its ringing frequency is related to its equivalent series circuit.

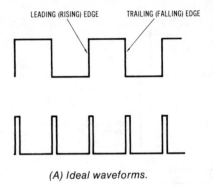

(A) Ideal waveforms.

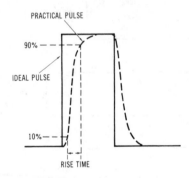

(B) Pulses with ringing.   *(C) Slow rise and fall.*

Fig. 19-7. Typical pulse distortion.

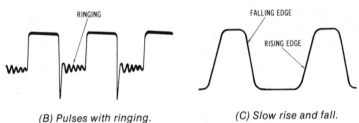

Fig. 19-8. Rise time of a pulse.

Ringing is of concern in a digital system if it occurs in a circuit that can be falsely triggered by the ringing undulations. In such a case, measures are taken to control the ringing and to reduce its amplitude sufficiently that system malfunction is avoided. One of the most common measures that is employed consists of added damping to the RLC circuit; in other words, a resistor with a value between 1 and 50 ohms may be connected in series with the input of the circuit that rings excessively. This

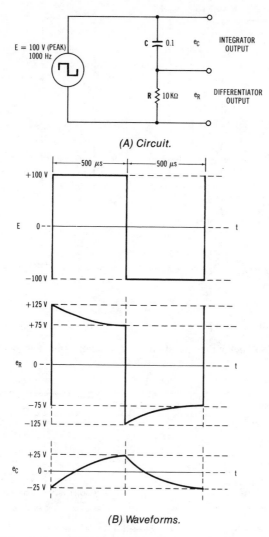

(A) Circuit.

(B) Waveforms.

Fig. 19-9. Action of RC integrator and differentiator on a square wave.

added damping involves a trade-off, in that the rise time of the digital pulse is slowed, and excessive rise time can also cause system malfunction.

The conductors (interconnects) in a digital system cannot necessarily be regarded as simple conducting paths with a small amount of resistance. Any conductor has at least a very small amount of inductance and at least a very small amount of dis-

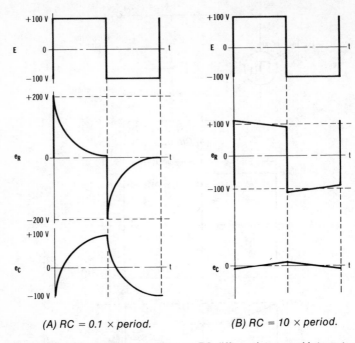

(A) RC = 0.1 × period.   *(B) RC = 10 × period.*

Fig. 19-10.   Effect of time constant on RC differentiators and integrators.

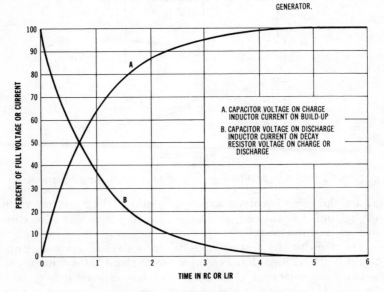

Fig. 19-11.   Time-constant chart.

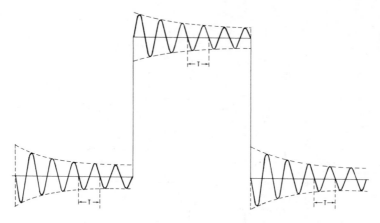

Fig. 19-12. Ringing transient response of an RLC circuit.

tributed capacitance. Digital pulses with short rise times have very high-frequency harmonics. These harmonics "see" the residual inductance and capacitance in a conducting path as well as its small resistance. Therefore, digital lines must often be regarded as RLC paths with respect to the high-frequency harmonic components in digital pulses. Improperly operated digital lines can impose overshoot, undershoot, ringing, or combinations of these basic pulse distortions. Because very high-frequency harmonic components must often be dealt with, even a comparatively short "run" of conductors on a pc board can be a potential source of digital-pulse distortion.

Since ringing is a persistent problem in digital system operation, manufacturers of gates often include input clamp diodes, as shown in Fig. 19-13. These diodes clamp the input lines to ground potential so that a line can drive the gate positive, but it cannot drive the gate negative. When a negative-going undershoot arrives, it is shunted to ground via the low forward resistance of the clamp diode. Although clamp diodes do not eliminate ringing waveforms, they assist in reduction of their amplitude and can prevent the generation of false signals.

BACKPLANE RINGING

The *backplane* is the area of a computer system where various logic and control elements are interconnected. A backplane sometimes has the form of a "rat's nest" of wires interconnecting printed-circuit cards in the back of computer racks or cabinets. Interfaces connected to buses often react as resonant

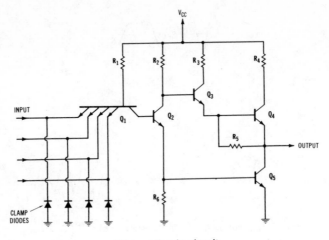

(A) Location in circuit.

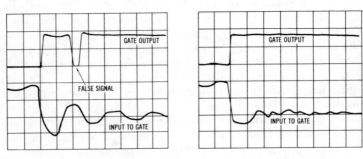

(B) Waveforms without diodes.   (C) Waveforms with diodes.

Fig. 19-13. Use of input clamp diodes to reduce transient ringing.

stubs (resonant line sections) that reflect pulse energy and generate ringing waveforms. Various expedients are employed to cope with backplane ringing. As shown in Fig. 19-14, if a line cannot be correctly terminated, it is helpful to provide a source resistance that is equal to the characteristic impedance of the line. If the source is matched to the line, reflected pulses from the load end are not re-reflected from the source end.

However, it is often impossible to match the load to the line or to match the source to the line. In such a case, ferrite beads are sometimes placed on lines as shown in Fig. 19-15. A ferrite bead provides a certain amount of reflection, and if it is strategically located on the line, it can provide some degree of cancellation of reflected pulses. A ferrite bead is effectively inductive

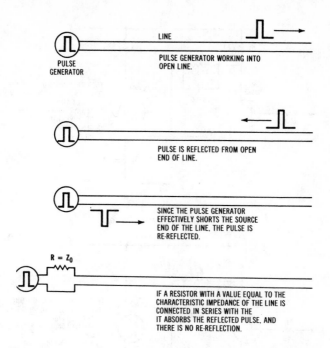

Fig. 19-14. Matching of source to line to eliminate re-reflection.

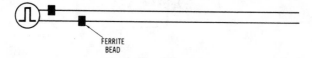

Fig. 19-15. Use of ferrite beads to assist in cancellation
of reflected pulses.

and can be compared with a small inductor. It can assist in the compensation of stub reactance.

Ringing is also reduced in some situations by the use of active terminators. An active terminator is typically a Schmitt trigger circuit; it exhibits a hysteresis action that helps to reduce the amplitude of ringing waveforms. A Schmitt-trigger circuit (Fig. 19-16) is a regenerative bistable configuration whose state depends on the amplitude of the input voltage. It is useful for waveform restoration and squaring nonrectangular inputs.

With reference to Fig. 19-16B, assume that Q_1 is nonconducting; the base of Q_2 is biased at approximately +6.8 volts by the voltage divider consisting of resistors R_2, R_3, and R_4. The emitters of both transistors are then at 6.6 volts, due to the

385

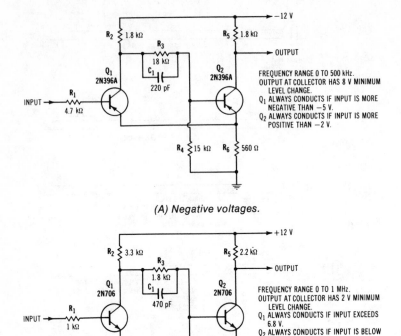

(A) Negative voltages.

(B) Positive voltages.

Fig. 19-16.   Schmitt-trigger circuits. (*Courtesy General Electric Co.*)

forward-bias voltage required by Q_2. If the input voltage is less than 6.8 volts, Q_1 is off, as was assumed. As the input approaches 6.8 volts, a critical potential is reached at which Q_1 starts to conduct and regeneratively turns off Q_2. If the input voltage is now reduced below another critical value, Q_2 will again conduct. The 1.6-volt difference between the response to an increasing or to a decreasing signal is called *hysteresis*.

Index